D1871562

ENERGY USE EFFICIENCY IN DRYLAND AGRICULTURE

ENERGY USE EFFICIENCY IN DRYLAND AGRICULTURE

R. MURUGESAN

KALPAZ

Energy Use Efficiency in Dryland Agriculture

Rs. 690

© R. Murugesan

ISBN: 978-81-7835-799-7

All rights reserved. No part of this book may be reproduced in any manner without written permission.

Published in 2010 in India by
Kalpaz Publications
C-30, Satyawati Nagar,
Delhi-110052
E-mail: kalpaz@hotmail.com
Phone : 9212729499

asser Type Setting by: Rudra Computer Graphics, Delhi
Printed at : Gian Ridhi Offset Printers, Delhi

Dedicated to my Parents
Thiru. K.R. Ramasamy
Tmt. R. Kannammal

Contents

Foreword 9
Preface 13
Acknowledgements 17
List of Tables and Figures 19
Glossary 21

1. **Introduction** 23

2. **Energy and Agriculture ; The Conceptual Context** 39

3. **Methods and Classification of Energy** 71

4. **Socio-Economic and Farmholding Practices** 91

5. **Energy Use Pattern in Dryland Areas** 111

6. **The Influence of Energy Input** 141

7. **Summary and Conclusion** 157

References 171
Appendices 195
Index 279

Foreword

Energy is a hot topic for discussion, particularly so in academics. The world has suddenly woken up to some startling facts - the energy reserves are not elastic; many of the known reserves are on the verge of exhaustion; the world will have to conserve energy in a major way in order to survive; energy conservation is not an exercise limited to factories and power stations alone; there needs to be a sense of energy conservation in all walks of life including farming; many simple and innocuous agricultural operations are much more energy profligate than what it would appear otherwise. It is this consciousness which has driven the author Dr R Murugesan to study the 'Energy Use Efficiency in Dryland Agriculture'.

The object of this study includes the existing cropping and energy utilization pattern of agriculture in dryland areas, the impact of energy inputs on selected crops therein and to make recommendations for sustainable energy management strategy. This study was conducted in one of the southern districts of Tamil Nadu and it includes operations like ploughing, harrowing, manuring: bunting or seed bed preparation, seeding or sowing or transplanting, fertilizer application, irrigation, weeding, pesticide application or plant protection, harvesting, threshing, drying, processing, packing, haystack making and transport of seeds, fertilizers and products. Surely, it is a wide range of agricultural operations that the study has covered.

Agriculture in the colonial era used negligible commercial energy and mainly depended upon human and animal energy sources. The Green Revolution brought in its wake HYV seeds, chemical manure, irrigation including canal irrigation from large dams and groundwater mined by electrical or diesel pumpsets and a thorough pesticide coverage. This led to a phenomenal increase in production and the productivity drew at more than 4 per cent during the decades of 70s and 80s. Food security was one of the main concerns before the nation. Along with the increase in productivity there grew facilities for all inputs supplies. It was only during the late 80s that there was slow down in agriculture and a plateauing of production. Strategies have been devised for 'putting the agricultural back on the rails'. Programmes like Rashtriya Krishi Vikas Yojana (RKVY) aim at intensive application of inputs for enhancing productivity.

Energy audit was never a concern of the Indian agriculture or those who managed it. It was only with the introduction of devices like pumpsets, tractors, chemical fertilizers and mechanized transportation that the question of energy efficiency within the agriculture cropped up. There is a substantial chemical fertilizer industry in the country mostly comprising urea units producing about some 24 mts of urea alone. Initially some of these units were coal based while others used naphtha for their feed stock. Natural gas based plants have now become the order of the day as the naphtha prices have sky rocketed. The urea manufacturing plants are rated according to their energy efficiency.

The Indian urea plants are amongst the most energy efficient in the world considering their age. This, however, is not a matter of solace. Whenever urea is applied in the fields there is a hidden energy cost to the system which translates into an economic cost on the nation. The urea and the phosphatic - potassic group of fertilizers are heavily subsidized. Hence, the true cost is not evaluated either in economic terms or in terms of energy.

The study brings forth clearly that the modern agriculture is highly energy intensive. Crops like sugarcane, for instance, consume more energy than most others, either in terms of fertilizers or in terms of irrigation and so is the case with paddy. On the other hand dryland agriculture does not consume much energy. However, this form of agriculture is in the process of giving way to what may be called the 'New Agriculture' using controlled conditions to cultivate crops like paddy and wheat. It is not that the consequences of this switchover from the traditional dryland farming are not known. Our agriculture scientists well acknowledged that the dryland crops like 'Ragi' or 'Bajra' consume much less water, much less energy and much less fertilizers as compared to paddy, wheat and sugarcane. In fact, these crops give more nutrients to the soil than what they draw from it. It is also well acknowledged that these crops have greater food calorific and, hence, food value as compared to rice and wheat. Yet, the macro policies of the State are such that the area under crops like 'Ragi' and 'Bajra' is on decline and crops like wheat and rice are heavily supported under our Public Distribution System. Thus, it is our macro policies which are responsible for promoting energy profligacy, as the study clearly finds.

Studies like this serve purposes more than one. To my mind, the study apart from focusing from need for energy audit of the agriculture sector also emphasizes the need for supporting alternative agricultural practices like permaculture, natural farming, diversified farming practices, integrating livestock management amongst others. It also advocates green manuring, composting and social harnessing. There is yet another purpose that the study serves: it jars us out of our sense of complacency.

BK Sinha
Director General
NIRD

Preface

Energy is the largest input contributing to the growth and development of the human race. It is the repository of resources for the development of human kind. The prosperity and well being of any field is evaluated depending on its energy supply and consumption. The demand for energy has been increasing day by day due to the aspirations of the ever-growing population. The level of energy demand requires the same level of supplies to meet this challenge. In order to cope with this challenge, it is necessary to meet our growing energy requirements in agriculture sector by economising the use of energy sources considering climatic conditions and aspirations of the people. In India, during the last *two* decades, the efforts were taken to create energy infrastructure to satisfy the requirements of commercial energy especially oil and electricity. Consequently, in plan after plan, the investment in large scale power projects has increased steadily, while investment for the benefit of vast section of the population to augment the energy supply has not been up to the level required by the agriculture sector particularly in the case of dryland cultivation.

Prior to Independence, the agriculture sector used very negligible amount of commercial energy. It mainly depended on human and animal energy sources. By and large the farmers used those inputs, which were raised on the farm and the production was also largely oriented towards self-consumption. With the advent of the Green Revolution. Indian agriculture have undergone a sea change due to adoption of high-yielding varieties of crops coupled with multiple cropping system, expansion of irrigation facilities and farm mechanisation which have encouraged the farmers to use higher doses of commercial inputs. The major component of this change was relative increase in the overall use of intermediate energy inputs. The changes have been both positive and negative: on the positive side a more stable and abundant food supply has resulted - on the negative side more environmental degradation, more dependence on fossil energy, and lower energy efficiency has been resulted.

It is increasingly becoming obvious that the conventional energy resources of India can not sustain even at the present rate of growth for a long time. In order to cope up with the vast energy demand of changing agricultural technology, a shift to the alternative source of energy might be beneficial in terms of increasing the availability of energy for rainfed agriculture and allowing a greater portion of the nation's effort to be devoted to raising the standard of living. In times to come, agriculture generally, would be required not only to produce more food grains but also to generate energy to cope with the increased demand.

Indian agriculture is something more than a mere economic activity. It is a culture, it ought to be conserved. Self-sufficiency in energy or sustainable energy in agriculture is a must for preserving the rainfed farming systems for the future generations to come. Here, the energy use in dryland agriculture is seen as a system and the energy analysis is done from a holistic perspective, seeing production, distribution, and consumption etc. as the parts of the agriculture. Since there is considerable inter-village and intra-village variation in the energy consumption pattern of our country, it is imperative to have a comprehensive micro-level energy planning for the maximum utilisation of the energy resources in general and non-commercial energy resources in particular. Naturally such a plan would lead to enhanced production resulting ultimately in balanced and sustainable development of the region. The need of the hour is to make a shift in policy framework from oil and electricity to agriculture energy and to have specific plans and budget outlays for it to meet the basic needs of the poor rural households / farmers in dryland agricultural regions.

Acknowledgements

Gratitude has always been latent dew on my mind and let me embellish this occasion to pour it out. At first, I wish to thank to God, without his grace I could not have completed my research. No words would be adequate to express my profound sense of gratitude and my heartful thanks to my guide and supervisor Dr.V.B.N.S. Madduri, Reader, department of Economics, School of Social Science, Hyderabad Central University, Hyderabad for his incisive and objective guidance of the entire research work. I shall remain grateful to him for ever.

I consider it as my foremost duty to express my special thanks to the authorities of the National Institute of Rural Development (NIRD), Hyderabad for given me permission to do this research during the course of work.

I shall be failing in my duty if I forget to express my sincere gratitude to Dr. T.C.A. Srinivasa Ramanujan IAS (Retd), the then Director General who has regularised my mindset in doing research.

I sincerely acknowledge the concerted assistance and professional advice of Prof. B. Sudhakar Rao, Director, Centre for Rural Industries and Employment (RIE) NIRD, in the submission of my thesis. I am glad to place on record my thanks to G. Ramachandraiah, Deputy Director, who rendered his precious time for going through my research chapters.

I am indebted to Dr. S. Ramaswamy, Professor of Economics, Gandhigram Rural Institute, Gandhigram for his moral support, valuable suggestions and kind encouragement. It is my duty to place on record my thanks to my friend Dr. M. Chinnadurai, Reader, Tamilnadu Agricultural University, Coimbatore, for his deserving help.

My parents and my wife deserve my deepest appreciation for their constant support, moral encouragement and inspiration. There are many others not named, who have helped me directly and indirectly with information and comments. To list them all would be impossible and to list a few individuals. A study of this nature which involved collective analysis of massive data would not have been possible without the willing co-operation and assistance of a large group of people from different villages of the study area.

Place: Hyderabad **R. MURUGESAN**

List of Tables and Figures

1.1	Share of Energy Consumption in India by Sector	29
1.2	Future demand for Commercial Energy in India	29
1.3	Category-wise All India Energy Input in Agriculture Production	30
3.1	Sample Villages and Concentrated Crops	74
4.1	Area profile – A Glance	93
4.2	Classification by Village Panchayats by Population Size	96
4.3	Number and Area of Operational Holdings	98
4.4	Area under Different Crops.	100
4.5	Sources of Irrigation	10
4.6	Agricultural Implements and Machinery	103
4.7	Age and Education of the Sample Farmers.	105
4.8	Landholding Pattern of Sample Farmers.	107
4.9	Sample Farmers' Livestock Particulars.	108

Figures:

Figure – I	Study Area	77
Figure – II	Location of the Sample Villages	
Figure – III	Irrigation on Cultivation	
Figure – IV	Classification of Land	

Glossary

ABE	-	Advisory Board on Energy
acr	-	Acre (100 cents)
BOF	-	Bullock Operated Farms
C	-	Celsius
CMIE	-	Centre for Monitoring Indian Economy
FAO	-	Food and Agriculture Organisation
FYM	-	Farm Yard Manure
G	-	gram
GNP	-	Gross National product
GOI	-	Government of India
h.m	-	human hour
ha	-	hectare (2.42 acres)
IREDA	-	Indian Renewable Energy Development Agency
j	-	joule
K	-	Potash
kcal	-	kilo calorie (4.184 x 10 joules)
kg	-	kilogram
kj	-	kilo joule (103 joules) (10)
km	-	kilometer
kw	-	kilowatt (10 joules/sec) (10)
kwh	-	kilo watt hour (3.6 x 10 joules)
l	-	liter
LF	-	Large farmers
LPG	-	Liquid Petroleum Gas

m.t	-	million tonne
MeF	-	Medium farmers
MF	-	Marginal farmers
MNES	-	Ministry of Non-conventional Energy sources
Ms	-	Mega joule (10 joules)
MT	-	Million tones
N	-	Nitrogen
NIRD	-	National Institute of Rural Development
nm	-	Nano meter
P	-	Phosphorus
PAR	-	Photosynthetically Active Radiation
q	-	quintal (100 kgs)
R&D	-	Research and Development
RET	-	Renewable energy technologies
SF	-	Small farmers
SPSS	-	Statistical package on social sciences
sq.km	-	square kilometer
TERI	-	Tata Energy Research Institute
Tj	-	Terra joule (10 joules)
TNAU	-	Tamil Nadu Agricultural University
TOF	-	Tractor Operated Farms

1

Introduction

1.1 Introduction

The practice of modern agriculture is becoming energy intensive in all aspects. In countries like India, where the area of cultivable land remains constant, the quantity of agricultural products should be increased to meet the needs of the growing population. This can be achieved by modern agricultural implements and fertilisers, which are highly dependent on non-renewable sources derived from fossil fuels like oil, coal, etc. With depletion of these fuels and escalation of the costs, it is now required to audit and conserve as much non-renewable energy as possible without affecting the agricultural productivity. It is also found that optimum uses of fossil fuel-based fertilisers will make future agriculture practice sustainable and eco-friendly.

To estimate the various energy input requirements of the agricultural practice with respect to different food and cash crops in the rainfed farming a field level study is conducted. In general, all crops depend on solar energy for photosynthetic activities, we are more concerned with the addition of commercial energy as inputs to enhance this activity. The cost of commercial energies keeps on escalating day by day resulting in the increase of cost of production. The patterns of agriculture production are also undergoing changes in various localities. For example, during the last two decades the energy use in agriculture was self-subsistence in nature and the villages were producing their own resources such as food for human, feed for animals and fuel for cooking. This sustainability in agriculture is slowly disappearing due to the introduction of cash crops like sugarcane, cotton, groundnut, coconut, tobacco etc. For real sustainability in dryland agriculture, the pattern of energy uses and crops should be optimised to make agriculture cost-effective, profitable and sustainable. The study aims at evolving this, for efficient usage of resources derived from non-renewable energy sources and appropriate rotation of food and commercial crops of dryland farming.

To achieve the above goal, T. Kalluppatti Block in Madurai district of Tamil Nadu is chosen. The statistical data, from the block-records with respect to cultivated area, cropping pattern, yearly rainfall, types of soil, water potential and population of drought animals are collected through a detailed questionnaire. The response from a randomly selected sample of 240 farmers is covered and data for the different types of fertilisers used, amount of labour input, kind of mechanical operations, cost of cultivation and other related information are collected.

With the knowledge of the energy content of various afore-mentioned agriculture activities the total energy inputs for various crops have been estimated. The energy input also varies with the level of agriculture holding. In this study, the farmers are classified into four groups namely (i) marginal, (ii) small, (iii) medium, and (iv) large farmers. This classification will show the amount of usage of machines, fertilisers, pesticides and electricity as well as the usage of draught animals. This will help us to find the ways to optimise the usage of machines and animals for efficient farming.

1.2 Energy Scenario of Indian Agriculture

The agriculture sector of the Indian economy prior to independence used very negligible amount of commercial energy. It mainly depended on human and animal energy sources. By and large, the farmers used those inputs, which were raised on the farm, and the production was also largely oriented towards self-consumption. With the advent of the Green Revolution in India, in the mid-sixties, Indian agriculture in general, the agriculture in Punjab, Haryana, Tamilnadu and Western Uttar Pradesh in particular have undergone a sea-change due to the adoption of high-yielding varieties of crops coupled with multiple cropping system, expansion of irrigation facilities and farm mechanisation which have encouraged the farmers to use higher doses of commercial inputs (Dhawan, K.C. et.al., 1993).

The major thrust of the Five Year Plan of the Government of India was towards the improvement of the agricultural sector. Indian agriculture experienced a significant change in the energy mix over the past three decades. One major component of this change was the relative increase in the overall use of intermediate energy inputs. The joint share of this energy inputs in the total farm production expenses increased from 22 per cent to 30 per cent during the period from 1960-61 to 1984-85. Moreover, this increase was accompanied by considerable shifts in the composition of their consumption (Kundu and Vashist., 1991).

The energy consumption in developing countries is only 17.5 per cent of the world's total energy consumption. In India, per capita energy consumption is increasing at about 2 per cent per annum which is almost equivalent to its population growth rate (Planning Commission: 1979).

In rural sectors, dry land agricultural farms have the potential to utilise the solar, biomass and wind energy. Utilisation of these resources in agriculture covers water pumping and solar drying (NSEC: 1983). There is a paradox of energy scenario in rural areas of developing countries. Either extremely little or too much energy is being used and the basic problems are to find ways to provide more energy and at the same time conserve energy (Ravelle, R :1978).

1.3 The Role of Agriculture in India's Energy Economy

India is engaged in planned agriculture development. In the last half century, the technological development of agriculture has dramatically changed the performance of both dryland and irrigated farming. The changes have been both positive and negative: on the positive side a more stable and abundant food supply has resulted; on the negative side more environmental degradation, more dependence on fossil energy, and a lower energy efficiency has been resulted.

Due to India's focus on the industrial sector in mid sixties and the increasing role of the service sector in the present past, the share of the agricultural sector in Net Domestic Product (NDP) has gone down significantly from 51.2 per cent in 1960-61. But it still remains to be the single largest contributor to NDP with a share of 32.6 per cent in 1985-86. Also, despite the rapid increase in urbanisation, nearly 75 per cent of our population continues to be dependent on agriculture for a livelihood. With such a large segment of the population being dependent on it, the agriculture sector as a whole has to ensure that, it:

- produces enough food for the people and raw material for industry at affordable prices.

- in the process of growth, it does not create regional imbalances.

- does not widen (in fact, it should reduce) disparity among the cultivators and between cultivators and farm labourers, and

- generates maximum possible gainful employment, thereby contributing to the alleviation of poverty.

A careful analysis, therefore, has to be carried out with the resources available to the agricultural sector. The two major resource requirements of the agricultural sector are land and energy. Land being more or less a fixed resource, agricultural productivity could be related directly to the energy inputs (Bose and Srivastava, 1988).

It is increasingly becoming obvious that the conventional energy resources of India can not sustain even at the present rate of growth for a log time. In order to cope up with the vast energy demand of changing agricultural technology, a shift to the alternative source of energy might be beneficial in terms of increasing the availability of energy for rainfed agriculture and allowing a greater portion of the nation's effort to be devoted to raising the standard of living. In times to come, agriculture generally, would be required not only to produce more food grains but also to generate energy to cope with the increased demand.

1.4 Energy Requirement of Agriculture Sector and Future Scope

"As agriculture in the developing world is in the process of substantial change, there is still scope for the introduction of energy-efficient technologies" (Edward Saouna : 1984). Both renewable and non-renewable energy sources play important role in improving the productivity of Indian agriculture. It has been observed by scientists that the quantity of fertilisers applied, quality of seeds sown, timely field operations performed in crop production system are some of the very important inputs which can improve the agricultural productivity in rainfed farms also. The following tables 1.1 to 1.3 also depict the energy use pattern and likely growth pattern of energy consumption during the last two decades of Indian agriculture. These figures indicate that the use of commercial energy is on the rise. Also the cost of these energies are rising every year. On the other hand, the non-commercial energy sources like 'Farm Yard Manure' 'animal energy', 'human labour' etc., are showing declining or stagnating trends in their use in Indian agriculture during the same period.

During the year 1980-81, the contributions of commercial and non-commercial sources of energies were almost equal, but the ratio is 2:1 by 1990-91 in favour of commercial energies. The ratio is likely to reach to 3:1 in favour of commercial energy by 2000 AD if the rising trend of agricultural productivity is to be maintained.

According to the estimates made by the National Commission of Agriculture, the population of India is likely to increase from 650 million during 1980-81 to about 935 million during 2000-2001 AD, making about 44 per cent increase during the period of 20 years. The estimated total food grains required at this future date will be about 69% in comparison to the base year. In view of this the per capita availability of useable land is likely to decrease from 0.29 hectare (1971) to 0.17 hectare (2000 AD), the only alternative left is to at least double the production per hectare, irrespective of dryland or wet land agriculture.

1.5 The Duel Nature of Agriculture

Agriculture must be compatible with both society's needs and the natural eco-system. Rapid population growth and the technical development of society have resulted in difficulties for farmers world-wide to maintain this duel compatibility. In fact, today farmers face demands for a high productivity as well as environmentally sound, sustainable farming practices (Giampietro and Pimental, 1993).

Table - 1.1

Share of Energy Consumption in India by Sector (in percentage)

Sector	*1953-54*	*1960-91*	*1970-71*	*1978-79*	*2000-01*
Household	74.6	67.4	56.2	51.2	23.1
Industry	12.1	16.1	20.7	21.8	46.5
Transport	11.4	13.9	17.5	17.9	20.9
Agriculture	1.0	1.5	2.5	6.0	5.6
Others	0.9	1.1	3.1	3.1	3.9
All	100.0	100.0	100.0	100.0	100.0

Note: Value for the year 2000-01 is a forecast.
Source: Planning Commission, Report of the Working Group on Energy Policy, GOI, New Delhi, 1979.

Table - 1.2

Future Demand for Commercial Energy in India

Source	*Unit*	*1991-92*	*1996-97*	*2009-10*	*2020-21*
Electricity	TWH	231	336	725	1300
Coal	Mt.	229	311	690	1345
Petro-products	Mt.	57	81.2	165	335
Natural Gas	bcum.	18.6	30.2	65	130

Source: Planning Commission, Government of India, Sept.1995-1996.

Table - 1.3

Category-wise All India Energy Input in Agriculture Production (Energy Input in TJ)

Input Energy Sources	*Forms and Type of Energy*	*Year* 1980-81	1990-91	2000 AD
Animate Energy (Human and Animal)	Direct non-commercial	116600 (14.85)	130750 (10.39)	145000 (8.3)
Seed and Farm Yard Manure	Indirect non-commercial	277245 (35.30)	286600 (22.77)	296325 (16.70)
Diesel and Electricity	Direct Commercial	39461 (12.66)	253075 (20.10)	454022 (25.5)
Farm Implements	Indirect Commercial	39349 (5.01)	57707 (4.58)	88775 (4.90)
Fertiliser & Chemicals	Indirect Commercial	252670 (32.17)	530682 (42.16)	796850 (44.60)
Total		785325 (100)	1258809 (100)	1780975 (100)

Source: Ojha (1993).
Note: Figures in parentheses are percentage to the total energy input.

In rural areas of developing countries, local environmental constraints like historically based techniques of production and socio-economic structures shape the agricultural activities. In other words, "subsistence farming system" are economically not sustainable when these societies interact with more developed socio-economic systems.

This enormous jump in farm productivity would not have been possible without large injections of fossil energy and machine power in the rainfed systems. In fact, the flow of energy input in modern U.S. agriculture is 50 times higher than in traditional agriculture (Pimental, D et. al., 1979). However, the higher income of modern farmers has a price: high-technology agricultural techniques depend on non-renewable stocks of oil and have negative environmental impacts, which lower the sustainability of the agro-ecosystem.

These impacts include soil erosion, reduced bio-diversity, and chemical contamination of the environment by fertilisers and pesticides, and mining of ground water. Hence, current intensive agriculture, with heavy technological subsidies of fossil energy, is ecologically not sustainable.

The farmers use more and more fertilisers due to the subsidy extended by the government. This extensive use of fertilisers affects the agro-ecosystem. Similarly, the subsidies in rural electric supplies have also become a major barrier in the spread of renewable technologies (Pachuri, R.K: 1993). Fertiliser subsidy has been taken as 50 per cent of the amount shown in government documents, the rest is deemed to be going either to the fertiliser industry or to the feed stock agencies (Gulati & Sharma, 1992).

1.6 Fossil Energy and Agriculture in Developing Countries

Modern techniques of farming in developing countries are based on massive injection of fossil energy. This results in lowering the energy efficiency (output-input ratios), and a rapid depletion of non-renewable oil stocks. The two forces driving this development are:

(i) increasing productivity per hour of labour of farmers (=increasing the income and standard of living of farmers, and making available more labour for other economic sectors/activities), and

(ii) the increasing productivity per unit of land area (= increasing the total food supply).

In spite of numerous negative effects in terms of environmental sustainability and energy efficiency with modern farming techniques, farmers in developing countries are adopting some of the modern techniques, especially, high yielding seed strains, fertilisers, irrigation and pesticides. The rapid growth of the world population is associated with the maximum expansion of fossil energy use, which has resulted in serious ecological imbalances (Abernethy, V: 1993).

The energy inputs use has increased up to 300 times. The existing direct commercial energy needs of 1142 X 10^9 MJ in 2000 AD to raise the agriculture production from 175 mt. to 240 mt. The non-conventional energy sources like animal dung, bio-mass, wind and solar are yet to be exploited in a big way. There is a big gap of 529X10^9 MJs in the year 1990 and 169X10^9 MJ in 2000 AD between energy demand and supply in agriculture production alone. If this gap is not bridged, the energy shortage will affect adversely the agriculture production in the year 2000 AD (Bhatnagar, A.P: 1993).

The successful integration of renewable energy technologies will help to improve the energy starved rural sector. However, there are unique opportunities and problems to be met with in the dryland agriculture. For instance, the economic appreciation and availability of renewable energy is very much dependent on the environment and climatic conditions of the area of installation as well as the season. The energy demand pattern of a particular location is viable over time and the types of energy end use (Castelli and Mazzetto, 1988). Thus, micro-level energy planning has to be done at village level, because there are many resources, technologies and alternatives (Monlik and Shukla, 1989).

In fact, all the energy related studies should help us to frame energy policies, chalk out energy plans and suggest optional energy mix. Especially, studies at micro-level are very much needed, because energy requirements for agricultural activities differ from area to area depending upon the environmental conditions and different agricultural practices. Therefore, many area-based studies are necessary to arrive at meaningful conclusions about energy use pattern in dryland agriculture. This book is one such study, that analyses the agricultural energy use pattern in twelve village panchayats in the T.Kallupatti block of Madurai District in Tamil Nadu. Further, it estimates the energy requirements of different farm sizes, with various operations of both irrigated and rainfed crops and determines some factors responsible for variations in the energy consumption level in dryland agriculture.

1.7 Energy in Dryland Agriculture

Agriculture is an art of converting the energy of solar radiation into fuel for consumption of human machine. Agriculture has very strong linkages with other rural sectors like households and rural industries in India. This linkage is a two-way process of demand and supply of energy to the economic system. Generally, energy requirement for agriculture includes direct energy inputs provided by human, animal and machine powers in land preparation and other formal activities related to crop production, irrigation, transportation of input and output material etc. It also calls for an indirect energy embodied in fertilisers, pesticides and machinery (Painuly, J.P. et al., 1992). Agriculture supplies energy in the form of non-commercial fuels such as crop residues, animal dung and wood to the households, feed to the animals and raw materials for agro-based industries (ABE., 1995).

The demand for energy in agriculture has increased considerably with the introduction of modern and efficient crop production practices which is expected to increase at approximately 8 per cent a year (each one per cent increase in production would require two per cent of additional energy input). In order to double the food production by 2000 AD, the food production in India has to rise by 3.7 per cent every year (Maheswari, R.C. et. al., 1986). In the developing countries, there are many constraints like availability of useful energy, ability to pay for the same and the resource of unskilled labour, the reason being the dwindling of agricultural lands and the increasing population. These constraints can be combated by a sustained use of renewable sources of energy to increase agricultural production in dry land areas.

Initiatives taken so far have changed the scenario of agriculture and have transferred it from traditional to technology oriented, from extensive to intensive, from subsistence to surplus, from mono-crop to multi-crop and from open cultivation to controlled cultivation. The commercial sources of energy utilisation has increased, which may finally lead to the degradation of the agro-ecosystem (Shama, R.N : 1992).

1.8 Perspective of the Study

Agriculture is something more than a mere economic activity. It is a culture, it ought to be conserved. Self-sufficiency in energy or sustainable energy in agriculture is a must for preserving the rainfed farming systems for the future generations to come. Here, in this study, energy use in dryland agriculture is seen as a system and the energy analysis is done from a holistic perspective, seeing production, distribution and consumption etc. as the parts of the agriculture.

Energy self-sufficiency and sustainability of the system is seen in a larger time frame. More than the economic feasibility or productivity maximisation, the thrust is given for viability optimisation. This is a requirement for sustainability of any system. Agriculture is a means for converting solar energy into usable forms for human beings and animals. It uses sunlight for doing photosynthesis. With the help of soil and water it makes food. The process is a continuous recycling of nature.

Any external input (fossil energy) in any rainfed agriculture system will make it unviable in the long run. In assessing fossil energy inputs and crop yields, it becomes evident that the marginal returns in increased production per input Kcal decreases as a particular input continues to increase, but at a decreasing rate, with each additional kilogram of nitrogen fertiliser applied until an input of about 200 kg per hectare is reached (Pimental,D., et.al: 1973). If the quantity of nitrogen applied increases further, corn yields often decline. As a result, the energy output/input also decline with the further increment of nitrogen becoming less favourable (Munson and Doll, 1959). To keep agriculture sustainable, the attempt here is to reduce external inputs to the possible minimum level.

1.9 Statement of the Problem

The economic conditions of agriculture and thus of agriculturists, mainly marginal and small farmers are worsening day by day. With the modern scientific understanding of soil and plants, we find the entry of the machinery, agro-chemicals, fertilisers (synthetic) huge capital and energy intensive (and of course, labour saving) farming techniques. The modern farming requires more commercial energy, which involves high monetary investments. It not only affects marginal and small farmers but also affects agro-ecosystem. In the long run, it can be a threat to sustainability.

For conservation of energy or application of eco-friendly energy resources for agro-ecosystem, there is a need for energy auditing in the existing farm practices of different categories of farming. Environmentally unsustainable crops and farming practices are coming up and they are tackled by energy assessment model and also supported through policy measures like utilisation of renewable energy (non-polluting) sources such as solar, wind, bio-mass and bio-micro-nutrient. At this juncture, there is an urgency to look into the problems of agriculture and work for an alternative energy management system for agriculture, which 'liberates' agriculture and make it sustainable. A holistic study of these issues is not yet made and thus the present study attempts to look into these issues.

1.10 Objectives

In the light of the foregoing probe, the following objectives are formulated.

(i) To study the existing cropping and energy utilisation pattern of agriculture in the dryland area.

(ii) To examine the impacts of energy inputs on output of selected crops of dryland and

(iii) To provide recommendations for sustainable energy management strategies that are needed for carrying out sustainable agricultural operations by using alternate and renewable energy sources.

1.11 Chapterisation

This book contains seven chapters. Following the introductory chapter, a review of earlier works in the field is discussed in second chapter. The third chapter contains research design, explaining sampling procedures, methods of measurement of variables and models used. This chapter also contains the definition of concepts used in the study. In the chapter four, a brief profile of the sample design is presented. Chapter five deals with the level of energy consumption of various crops and different categories on a per acre basis. A regression model to explain the major influencing energy inputs for the selected crops with different crops with different categories are discussed in chapter six. The last chapter summerises the discussion draws conclusions, offers policy suggestions and indicates the possible line of further research.

1.12 Summary

Energy is the largest input contributing to the growth and development of the human race. It is the repository of resources for the development of human kind. The prosperity and well being of any field is evaluated depending on its energy supply and consumption. The demand for energy has been increasing day by day due to the aspirations of the ever-growing population. The level of energy demand requires the same level of supplies to meet this challenge. In order to cope with this challenge, it is necessary to meet our growing energy requirements in agriculture sector by economising on the use of energy sources considering climatic conditions and aspirations of the people.

2

Energy and Agriculture : The Conceptual Context

2.1 Introduction

Several concepts relating to the subject are being spelled out in order to have a comprehension of the problem as such. The relevant concepts considered for the purpose of the study are presented as under.

2.2 Meaning and Definition

The term 'energy' has been derived from the Greek word 'energia' meaning capacity or ability to work. The evaluation of civilisation in terms of the invention of tools and weaponry, the development of agriculture and animal husbandry, and acquisition of other aids to living is just another facet of how much, in what form and in which way man is used to energy. Although there is no substitute for energy, it has no value in itself except through the functions that become possible by its use (OECD, 1975). So, Energy even in physics is clearly defined as the capacity to work and power as the rate at which the work is done.

William Blake, an English Poet said, 'Energy is the only life....... Energy is eternal delight'. It can be said that man's activities basically depend on energy. This dependence may be direct and visible as in the case of heating for food preparation or indirect and invisible, for example, in the production of corn (Simon, A.C.1975). It is in this sense, energy is deemed to impose a production frontier on economic activity and comprehend to be a potential limiting factor to economic growth. Dorf (1978) defined energy as anything that makes it possible to do work and capable of bringing about movement against resistance.

All matters and all things have energy. Therefore, energy is a property of matter. It takes many forms and it can neither be created nor destroyed. When energy is used it only changed into another form (Dorf, R.C.1978).

According to Dayal, M (1983) old Indian analysis of ***'Shakti'*** dealt with energy as creating matter and being created from it, changing and appearing in different forms and essential one with manifestations in different forms and areas. Some similar concepts of energy have been arrived at by the discoveries of modern theoretical and experimental physics including the interchangeability of mass of energy and the implied theory for the different basic forces of nature as known to physics.

The different forms of energy used by mankind are derived from sources like fossil fuels, bio-mass, flowing water, wind, solar radiation, nuclear energy, etc. Even though the ultimate source of almost all forms of energy is seen except nuclear energy.

Sathish Chandran (1977), classified energy into flow resources and stock resources or renewable and non-renewable. Energy from plants, animals, sun, wind and water are renewable, whereas energy obtained from fossil fuels such as petroleum, coal and natural gas are non-renewable.

According to Crambie Aber (1978), Commercial energy either involved a technology of energy conversion or entered commercial channels while non-commercial energy is consumed directly without technological conversion and even without being formally marketed.

Venu (1980) classified energy into 'income forms' and 'capital forms'. Income forms of energy such as hydro, solar, geothermal, tidal and wind power and produced continuously in nature is thus replenishable. The later i.e., capital forms of energy are accumulated over time which are theoretically not readily replenishable and the stocks are finite at a given point of time. He also considered fossil hydrocarbons, nuclear energy and hydropower as commercial energy sources, which involve capital intensive technology. Non-commercial energy, which included firewood, dung, vegetable materials, and human waste, caters to the requirement of rural and unorganised sectors of the economy.

2.3 Energy in Agriculture - A Review

It is necessary for any researcher to gain knowledge by pursuing the related literature and the views expressed by various experts. It helps to conceptualise the factor to be studied and to formulate sound methodology. An attempt has been made in this chapter to review some of the previous studies related to the present work.

Studies on energy in agriculture can be divided into three groups based on their scope:

(i) Studies relating to energy consumption for managing livestock;

(ii) Studies covering energy consumption for crop production (pre-harvesting energy) and

(iii) Studies on energy consumption for post-harvest operation including transportation of agricultural produces.

The studies carried out have covered any one or two or all these three aspects. As the present study focuses its attention on energy consumption for rainfed crop production and their processing (post-harvest operation), an attempt is made to review the studies covering agriculture energy issues, which fall under two categories viz.

(i) Studies covering non-commercial and commercial energy for pre and post harvest operation;

(ii) Studies covering dryland agriculture energy use as a whole with relation to agro-ecosystem.

This review of literature is presented on the basis of these objectives. The studies on the energy use in rainfed agriculture in general, and also on commercial and non-commercial energy aspects are presented first. The studies dealing with the energy requirement for different farm sizes, various mechanisation levels and the energy conversations are presented next.

2.4 Energy Analysis in Dryland Agriculture

Even in this 'modern' world, majority of the people directly and others indirectly are depending on agriculture for their survival. In a highly populated country like India with comparatively high density, agriculture is and will be the single largest occupation. In nearly every country, total food production has increased in recent decades, because the development of agriculture has altered the structure and composition of selected systems so as to increase their yield of products. However, despite their modification by cultivation, irrigation, animal and fertiliser use and other management techniques are all not controlled by man, but by energy.

At the end of the decade, mankind was facing a pressing need, not only to raise the production but also to make distribution, more effective and to improve the supply of energy.

The trend towards increasing intensive dryland agriculture also involves substantial changes in fertilisers and pesticide use, which in turn requires considerable energy inputs (F.A.O., 1975).

Farm power and farm energy are two distinct terms, which in literature have been interchangeably used. Technically, power refers to the rate of doing work and energy to the product of power and the duration of time for which the power is supplied. A real measure of the mechanical component of farm technology is energy input since it differs from region to region. Normally, four common sources of energy viz., human energy, animal energy, mechanical energy consisting of oil based and electricity based machines, and the Chemical and Fertiliser energy are being considered for any agricultural research.

In the co-ordinated scheme for Energy Requirement for Intensive Agricultural Production operated in Coimbatore, Pantnagar and Ludhiana (in 1973-74), and in a study conducted by Girija Sharan et al.,(1994), these types of farm energy have been identified.

Kahlon (1994) observed in the co-ordinated scheme report from Punjab Agricultural University that, in tractor operated farms, tractors are the main sources of oil engines and are the only farm of mechanical power. Spedding and Walsingham (1975) commented that agriculture has always been man's main way of harnessing solar energy and it is likely to remain so far a long time.

Energy has been termed as the fuel of economic progress. It is the prime mover of economic growth and development (Smil et al., 1980). Next to household, agriculture accounts for the highest consumption of energy in rural areas. It is seen that energy input per unit production is much higher in developing countries than the developed countries. The reason being that the efficiency of the whole farming process is lower.

The overall efficiency of energy use in agriculture in India is only 6 per cent while in developed countries it is between 25 to 36 per cent. There is however, one advantage i.e., additional kcal used in the traditional modes of agriculture are obtained from local sources in the form of agriculture are derived from scarce petroleum products (Bansal, N.K: 1988).

The population pressure, food requirements, agricultural yields, agricultural energy use pattern and environmental condition are all interrelated irrespective of irrigation and dryland conditions. That is why many research studies have been taken up all over the world to scientifically analyse energy and agriculture. For instance, Tsatsarelis (1992) has analysed the energy productivity, energy intensity and energy efficiency and reported different values. These differences are mainly due to different production systems, machinery, inputs used, and crop yields. Black (1971), made a study of five crops grown in the U.K. but included only tractor fuel in his measure of the energy subsidy. Perelman (1972) compared U.S. food production with Chinese wet rice cultivation and found the latter more efficient. Probably one of the most influential agricultural analysis was made by Pimental et al. (1973). They analysed the changes in U.S. maize production over a quarter of a century and showed that yields have shown an increase of 138 per cent at the cost of a 310 per cent increase in fossil fuel expanded in production.

The energy ratio is the ratio of the calorific value of a food to the energy required to produce it. Black (1971) first applied this ratio. Fluck (1979) reviewed the utility of this measure and its limitations.

Steinhart and Steinhart (1974) estimated the energy used in U.S. food system considering the energy used in the farm, in the processing industry, and in commercial establishments and home. They reported that of the total 2172 10^{12} kcal of energy used in United States' food system, about 526.1 10^{12} kcal were used in the farm and the remaining quantity in the processing industry, commercial establishments and home.

Quantifiable relationships have been demonstrated between the energy expanded in food production and the food produced. Slesser (1973) studied 131 food production systems and was the first to show that the food output was correlated to the energy input. The study also documented the applicability of the law of diminishing returns in dryland farming, at higher energy intensities. Yield responses to increase in energy input are small. Edwardson et al., (1977) had a similar findings from data over 350 systems, where they compared the protein output to the gross energy requirement. It is thus, necessary to make detailed energy-use analysis in order to evaluate alternatives for dryland areas.

The conventions for such analysis were reported in IFIAS (1974). Chapman (1974) and Pimental (1974) provide a review of the methods. Typically an energy system diagram is used to summarise the main components and pathways and as means of assembling data on energy and material flows. Several authors (Fluck and Baird (1989); Leach (1975); Mitchell (1979); Odum (1967, 1971, 1983); Stanhil (1978, 1981) have endorsed this approach.

Swaminathan (1977) advocated that the efficiency of the production technique in the rainfed areas should be measured by energy input-output ratio in addition to yield and income.

Researchers from various countries, calculating energy demands for agriculture production as well as energy productivity, intensity (e.g., energy input per unit of output kWh/kg), and energy efficiency (the quotient of energy output and the energy input) reported different values. This is due to different production systems followed by the farmers (Tsatsarelis, C.A. 1993).

Rajasekhar and Poyyamoli (1993), found that the total output from all sources is 19206.73 kWh/ha. Among the total input energy, 93.43 per cent of the input is from commercial sources, while 6.57 per cent of the input is from non-commercial sources. The inputs, ranked source-wise are, agro-chemicals (87.01%), electricity (6.71%), draught power (2.81%), human labour (man) (2.68%) and human labour (women) (0.79%). The yield is 68 tonnes/ha and the residue 14.96 tonnes/ha giving an output to input ratio of 20.42.

The total energy requirements for production of sugarcane-sugarcane (ratoon) rotation were 32825.26 to 42083.67 kWh/ha and 20805.72 to 20861.28 kWh/ha at Ludhiana and Pantnagar centres respectively. Therefore, at Ludhiana centre more energy is required for the crop production in case of sugarcane-sugarcane (ratoon) rotation. This was primarily attributed to higher doses of fertilisers and water use at Ludhiana centre. The energy output - input ratio at Pantnagar centre (15.16 to 16.04) was almost 2.8 times more than the ratio (4.54-5.78) at Ludhiana centre (Mittal, V.K. et al, 1983).

Stout (1981) assembled data from FAO statistics and showed that in developed countries the average crop yield was 2631 kg/ha with 5388.93 kWh/ha (19.4 MJ/ha) of energy input and in developing countries yield averaged 1328 kg/ha with one-tenth as much commercial energy input. Makhijani and Poole (1975) analysed the data on energy requirements of agriculture for India, China, Taiwan, Japan and USA are found that as the uses of irrigation and fertiliser increase the production per hectare and that the total energy requirement per ton of the crop decreases rapidly.

As stated earlier, the energy consumption pattern for the same crop in different parts of the country in the same season is not the same. Variability is very large and it is due to agro-climatic and soils factors. Among the different segments of energies used in crop production system, the method of irrigation and sources of irrigation water, play a significant role in arriving at the total value of energy use. It appears that the efficiency of the system makes a significant effect (Ojha, T.P. 1993).

Palaniswami (1975) observed that among the food crops, energy requirements for high yielding variety rice was higher than that of millets like sorghum, bajra, ragi. For rice, the energy consumed for irrigation alone took 65 to 68 per cent of the total energy followed by preparatory cultivation, since irrigation requirements were very high in terms of quantity and frequency. Srivastava (1984) shows that in Indian conditions, tillage and sowing operations (including fertilisers) consume 50-60 per cent of the total energy used in agriculture.

Kamble and Panwar (1986) found that tillage required maximum energy followed by irrigation, harvesting and threshing for seed production in IARI at Karnal for all kinds of crops.

Rajendran et al., (1987) pointed out that 4640 kWh/ha energy was consumed in the paddy cultivation of which 3.2 per cent was by human energy, 4.5 per cent by bullock energy, 0.05 per cent by mechanical energy and 92.25 per cent by electrical energy.

Mittal et al., (1985) presented the results of various field surveys of the AICRP energy requirement studies on the energy use pattern in various parts of the country. Sinha (1986) discussed the energy balance in agriculture in the developing countries with reference to India. Data on the energetics of rice and wheat production are also presented. Ramesh (1991) reported on the source-wise and operation-wise details of various crops grown in Pondicherry.

Sheshadri (1992) documents a decrease in efficiency of the Indian agricultural system from about 4.0 in 1951 to 1.25 in

1987. This is attributed to an increase in inputs to the tune of 900 per cent while output shows only a three-fold increase.

Some studies on agricultural energies are: Dekkers et al (1974) - the Netherlands; Pimental (1977) - USA; Zuchetto and Jansson (1979) - Sweden and Stout (1982) - Senegal. Makhijani and Poole (1975) presented an account of third world agriculture.

2.5 Farm Size

Thakur et al (1976) found that the total energy applied per unit area of seed potato, ginger and hill capsicum decreased with the increase in the farm size. The gross income per hectare and gross and net income per unit of energy were observed to be highest in the case of tomato followed by ginger and was the least in the case of green pear.

Singh and Singh (1976) reported that tillage operation consumed the bulk of energy used in the production of crops followed by irrigation and threshing. They also found that bullock operated farms used substantial portion of their energy requirement in the form of mechanical energy in the execution of irrigation, threshing and partly for land preparation through hiring of tractors. The energy use per hectare of operated area in tractor operated farms was found to be more than double of that used in bullock operated farms and approximately nine times of the energy content was supplied through mechanical source.

Sukhla et al., (1976) conducted a study in different areas of Bareilly district of Uttar Pradesh and reported that the use of electricity and tractor power was higher in developed areas and also on large farms.

2.6 Energy Sources

The energy sources in general are classified as commercial and non-commercial in these studies.

Yanagida and Conway (1981) stated that the measurement of total energy accounted not only direct energy use associated with fuel for propulsion and utilities but also indirect energy embodied in factors such as fertilisers and chemicals, tractors and barns.

Pathak (1983) classified the energy used in dryland agriculture into commercial and non-commercial. Commercial energy comprised of electricity, diesel and equipment, fertilisers and chemicals that were obtained through commercial channels whereas non-commercial energy included Farm Yard Manure (FYM), seed and draught animals. He also grouped energy sources into direct and indirect, the former included the energy supplied by human labour, draught animals, oil engines, electric motors, power tillers, tractors, etc., whereas the latter consisted of the heat content of the input or energy used in the production of fertilisers, chemicals, seeds and equipment.

Bisaliah (1985) gave a brief outline of trends in energy consumption pattern in India. Machinery and fertiliser together claim about 95 per cent of the total energy used in agriculture in the world. Energy intensive inputs such as fertilisers, pesticides, pumpsets and tractors have been emerging as important inputs in Indian agriculture. He suggests that it is better to stop using commercial energy in agriculture. There are a number of technological options such as production of plant nutrients from weeds, use of bio-gas for operating agricultural implements and machinery and use of solar and wind energy for operating irrigation pump sets. However, technological design in all these areas is not matched by economic design.

Irrigation consumed the maximum energy in all the farm operations in paddy (81.9 per cent) and wheat (38.1 per cent). The output-input ratio for paddy-wheat rotation was 4.59. The specific energy consumption for paddy was 5.87 MJ/kg, and that of wheat was 4.46 MJ/kg. Paddy consumed 90.9 per cent and wheat 94.3 per cent of the total energy input from commercial sources and the rest of the energy needs was met through non-commercial sources. Keeping in view of the fast depleting commercial energy sources, on alternative crop rotation may be thought which requires less energy in the form of fertiliser, as well as irrigation, and may give equivalent output-input ratios if not more. Introduction of oil crops and pulses on a large scale may be alternative crops to reduce the load on commercial energy sources (Surendra Singh, et al, 1990).

Mittal (1992) has calculated the total energy consumption and its distribution among commercial and non-commercial energy sources. He estimated the total energy values of sugarcane crops as 10763.3 kWh/ha (100%) of which the commercial and non-commercial energy contributions were 9463.89 kWh/ha (34070 MJ/ha) [87.9%] and 1299.4 kWh/ha (4678 MJ/ha) [12.1%] respectively. The contribution by seeds to the total energy value seems to be the maximum.

2.6.1 Commercial Energy

A study by Pathak, B.S. (1985) revealed that agricultural production increased by 175 per cent during the period 1965-66 to 1979-80; total energy consumption increased by 387 per cent; and commercial energy consumption increased by 1225 per cent.

Mitchell (1979) discussed the Indian agricultural system. He showed that 9 units of food energy were harvested for every unit of fossil energy put into the system.

Fertiliser, pesticides, farm machinery and irrigation respectively use about 60 per cent, 30 per cent, 7 per cent, and 3 per cent of energy. Increased use of modern techniques and to increase the commercial energy consumption and a careful planning is required to optimise the use of these items in dryland agriculture (Bansal, N.K., 1988).

Chemical Energy

The consumption of fertiliser nitrogen (N) in the crop year 1990-91 was 8.0 million tonnes and is likely to increase to 12-14 million tonnes by the year 2000 A.D. (Tandon HLS, 1991).

Singh and Govil (1968) explained that the optimum yield predicted for 1967-68 was 44.1 q/ha for a fertiliser combination of 180 kg of nitrogen and 106 kg of P_2O_5/ha, while the optimum yield for 1968-69 was 59.7 q/ha for a nitrogen application of 109 kg/ha plus 56 kg of P_2O_5/ha.

A long-term experiment was established in 1979, to find out the responses of sorghum to potash (K) application. Grain yield of sorghum was 4 tonnes/ha with an application of 120 kg of N/ha, 4.6 tonnes/ha with 120 kg of N plus 30 kg of K/ha and 5 tonnes/ha when 120 kg/ha each of N and K were applied (Prasad, R. et al.,1992).

In traditional agriculture there were heavy losses due to crop diseases as there were no scientific measures to check these diseases. But one positive aspect of traditional agriculture was that the local (deci) seeds were mostly disease residents. On the other hand, High-Yielding Variety (HYV) seeds, which are being used now, are prone to a number of crop diseases. That is why HYV seeds cannot be grown without insecticides and pesticides.

However, there is a negative aspect of the use of insecticides and pesticides. They are considered as a major source of pollution not only for plants, but also for human beings. According to Praful Bidwai (1982), large tracts of agricultural land in different parts of India are on the brink of the 'pesticides treadmill' and many districts are already in its grip. The pesticides treadmill is, ecologically, the most dreaded and economically, the most injurious phenomenon produced by the indiscriminate use of toxic agro-chemicals. In Gujarat, cotton farmers sprayed their fields five to eight times, ten years ago. Today they spray them twenty to thirty times with more potency ecologically more irrational and expensive pesticides.

In traditional system of agriculture, farmyard manure and green manure were used to increase the yield of the crops. However, the increase in yields was not significant as a result of these manures. Research in the field of agriculture led to the discovery of chemical fertilisers, which could fulfil the varying needs of the soil. As a result of this research, chemical fertilisers were prepared which could meet deficiencies of the soil in various nutrients. Some soils needed more nitrogenous fertilisers while others needed more of potash or phosphate fertilisers. The use of these fertilisers helped in overcoming the deficiencies of the soil. As a result of this the yields of crops increased considerably.

Parthasarathy (1973) conducted a study in Andhra Pradesh and found that for both the local and improved rice varieties the yield stops growing after 100 kgs/ha of nitrogenous fertiliser intake, giving an output of 39.38q/ha and any further increase in fertiliser application actually produces a decline in yield.

Biswas et al., (1987) analysed that the fertiliser recommendations need to be based on soil and crop response correlation studies. It can provide basic understanding of the residual effects of applied nutrients and effects of succeeding crop. They also suggest that suitable dose of fertiliser for black gram and green gram could be around 20 kg of P_2O_5 and 20 kg of P_2O_5/ha is of higher magnitude than the response to N both under rainfed and irrigated conditions. For groundnut, 20 kg of N_2, 60 kg of P_2O_5 and 20 kg of K_2O/ha appeared to be optimum.

Mechanical Energy

A common denominator to all developing countries is that they have been able to construct, either indigenously or with external sources, technology that has increased the productivity of labour and domestic resources. Dryland agriculture is no exception and the mechanisation of farming can have major effects on crop yields and agricultural productivity. For example, one can compare maize yields for mechanised agriculture of 5394 kg/ha and 5020 kg/ha for the USA and UK respectively, with yields of 1066 kg/ha for bullock operated farming in Guantemala and 1004 kg/ha for farming in Nigeria using human labour alone (Pionental and Pinental, 1979).

Farming systems were defined differently based on various criteria such as land holding structure, organisational set up, nature of ownership, cropping pattern or crop rotation, irrigation sources and degree of mechanisation.

Newbury (1980) described land holding structure, land use, its organisation and mode of application with regard to labour. The operational skills and techniques employed and all the ways in which capital was employed in terms of plant, machinery, technical aids, seeds and livestock etc. Johnson (1979) and Sivakumar (1982) identified three common sources of energy that were available for farm operations viz., human energy, bullock energy and mechanical energy.

Singh and Miglani (1976) considered irrigation energy, human energy, bullock energy, tractor energy and stationary machinery energy to find the energy requirements in Punjab agriculture.

Sharan et al (1975) referred to human labour and bullocks as biological sources and tractor as mechanical source. Sonerdon (1977) referred energy input in agriculture, as external energies required for machines, irrigation, fertilisers, seed production, and artificial crop drying and similar activities.

Black (1971) was one of the firsts to apply energy ratio, which he termed "Efficiency ratio" to industrialised agricultural production systems. He concluded that industrialised agriculture was no more efficient in energy-use than hand or draught animal cultivation. TNAU (1981) revealed that total energy required in bullock farm was found to be a maximum for ragi among short duration crops and was least for maize. In the case of tractor farm, sorghum required the highest energy followed by groundnut and maize.

Shyam and Nagaih (1982) in their energy analysis of potato production revealed that the major share of energy input is contributed by nitrogenous fertiliser and diesel in the case of tractor operated farms and seeds in the case of bullock operated farms, have been found to be the second largest energy input.

Singh, L.R. and Singh, B. (1976) estimated that the energy consumption for paddy [HYV] was 736 kWh for bullock farm, 883 kWh for tractor operated farm.

As a result of the technological advancements all over the world, the agricultural sector has also progressed. Agricultural machinery was introduced to replace, and in some cases, to supplement animal and human power machinery was introduced in the rural communities of India after independence. The most important machine, which is used in modern agriculture, is the tractor.

From the study of Palwinder Singh (1992) it is clear that out of the total landholders in the study area (Punjab), about fifty per cent owned tractors. Those respondents, who did not own tractors, hired them from the tractor owners. Dhawan and Singh (1976) studied the energy requirement at different levels of technology in Punjab. They indicated that in the existing situation human labour use was less on the tractor-operated farms as compared to the bullock-operated farms. Also there was a sharp decline (90 per cent) of energy supplied by the animal sources in the tractor-operated farms.

Singh and Chancellar (1975) presented a detailed study of energy inputs in different farm categories, namely bullock farms, tube-well farms and tractor farms of Meerut district of Northern India. They concluded that there was little evidence to show that the type of power source used by the farmers could affect crop yields and that increase in yield was the integrated effect of all inputs. However, the availability of mechanical power for higher rates of applications during specific periods permitted farmers to use different production strategies thus resulting in the increased annual production of food and commercial crops.

Irrigation

Traditionally, most of the land was unirrigated. Moreover, the irrigation was dependent upon rainfall in the study area. But with the introduction of agricultural technology and new high yielding varieties of seeds, the new irrigation sources in the form of electric pumps, diesel engines, canals and dams came into existence. Though this study concentrates on energy use in dryland agriculture, there is a necessity to compare with the energy utilised in irrigated farming system for better understanding of the energy use.

Aulani and Chancellor (1977) worked out the energy requirement for wheat production and use in California by including the energy use in the farm, transportation, processing, distribution and home. They concluded that total energy consumption for production and use of one unit of wheat was 1.7 times the calorific value of wheat. The energy ratios were higher for non-irrigated wheat than irrigated wheat.

With the advent of Green Revolution, the role of energy in raising the yield levels through energy utilisation and irrigation assumed greater significance and with this, emphasis on rural electrification has also changed from more village electrification to pumpset energisation (Rao and Bhavanishankar, 1986). Dvoskin et al. (1977) worked out the energy required for applying one acre foot of water in the 17 Western States of U.S. and found that on an average it took 1,134,660 kcal of energy to pump one acre foot of water from an average depth of 196 ft. to ground level.

Larson and Fangmeier (1977) observed that irrigation required most of the energy with groundwater sources and fertiliser was the second largest input with crops requiring nitrogen.

Patel and Pandya (1979) estimated that the total energy requirement for irrigation pumping in India for different periods taking into account human, bullock and mechanical energy sources. They found that the energy requirement increased from 5295 million kWh in 1968-69 to 11861.7 million kWh in 1978-79 and it was estimated that the requirement would be 18722 million kWh in 1998-99.

The study conducted by Rajendran et al. (1987) at Coimbatore revealed that under gardenland condition electricity was the major source of energy for crop production. It constituted more than 75 per cent of total energy input in crop production because of energy intensive lift irrigation. Electricity consumption was highest for banana followed by sugarcane and turmeric because of the long duration of these crops.

If we add new irrigation capacity to the tune of 35 million hectares by 2000 A.D., and adopt the norm of 1.5 tonnes for each additional hectare brought under irrigation, the added grain output would be about 53 million tonnes. To this be added 22 million tonnes that may accrue from a rise in irrigated yield in the existing 45 million hectares under foodgrains. Together, the irrigated farming can thus meet out our food needs by 2000 A.D. (Dhawan, 1988). In the present study, irrigation has been recognised as one of the most important agricultural input because the use of other agricultural inputs like chemical fertilisers, pesticides, seeds, solar energy etc. depends on the availability of irrigation.

2.6.2 Non-commercial Energy

Animal Energy

The use of animal power and machinery have been taken together as they are sometimes used as substitutes and at other times are used to supplement each other. No doubt, before green revolution, India's farming was dependent on human and animal energy. Like other developing countries, the animal power is an inseparable part of agriculture in India.

Animate energy includes human and animal energy. The method of measuring energy input for the human is to measure the metabolic energy required carrying out certain farming tasks. On an average, the energy value calculated for human and animal labour for work in crop production is about the same as the energy value for food consumed by the labour (Pimental, D. 1979). Unfortunately, we have no single accepted system for measuring the energy input for labour service to agricultural systems. Clearly, this depends on the objective of each study.

Macmillan (1987) has sorted out the problems in the measurement of draught animal performance. It is highly variable, both in time and place, and is subject to many constraints of a physical, physiological and behavioural kind (Brody, S. 1945).

The maximum pulling capacity of an animal is decided by many factors, e.g., species, breed, sex, size, age, health, nutrition, training, work posture, ground conditions and other wide range of variables which determine its working conditions (Pathak, B.S., 1987).

In agriculture based countries like India, energy for farm operation is designed mainly from animals. Animals are being used for draught purpose almost since the beginning of civilisation. It has been found that each bullock is capable of generating on an average, 0.373 kWh. The draught animals in India generate 30 million kWh of electrical power. The power output is also influenced by climatic conditions, nature of work and implementing efficiency. The animals are capable of providing maximum pull of 15-20 per cent of their body weight (Singh, S.N. et al, 1991).

Gupta et al., (1987) have studied the power requirements for rainfed agriculture in Harikot village of Hissan District in Haryana State. This study is purely based on human and animal labour requirements during the peak months of October, November, April and May. And also the study discussed the animate energy required for producing crops on bullock power-oriented farms.

Rajendran and others (1987) analysed the human and bullock energy utilised for irrigated crops and dry crops in the farms of Coimbatore District of Southern India. Among various crops they studied, sugarcane consumes more human labour energy (337 kWh/ha) and maize consumes less (86 kWh/ha) in the irrigated farm. In case of bullock energy, banana consumes more (218 kWh/ha) but less energy is consumed by sugarcane (ratoon) (12 kWh/ha). In dryland both human and animal energy utilisations were very less compared with irrigated land.

Farm Yard Manure (FYM)

Traditionally, organic sources of nutrients like Farm Yard Manure (FYM), compost and crop residues have been important inputs in crop production for maintaining soil fertility and to ensure yield stability.

De Wit, (1979) divided the energy used in farming into direct and indirect sources and used the term 'added energy' to refer to the sum of direct and indirect energy used in farming.

Rasmussen and Nielson (1978) divided the total energy consumption in the agricultural sector into primary and secondary consumption. Primary consumption consisted of petrol, oil, gas and electricity directly used in production whereas secondary consumption included energy embodied in fertilisers, pesticides, food stuffs, machinery and other inputs the manufacture of which demanded great amount of energy.

Hegde et al., (1992) suggested that 12 tonnes of FYM/ha could substitute 40 kg of N/ha of chemical fertilisers in rice. Further, they observed that there was residual effect equivalent to 30 kg of N and 30 kg of P_2O_5/ha as chemical fertilisers on the yield of succeeding wheat, so that, the cumulative effect of 12 tonnes of FYM/ha was equivalent to 70 kg of N and 30 kg of P_2O_5 in rice-wheat cropping system.

The use of FYM not only helps in substituting a part of fertiliser needs but also improves the general productivity by increasing soil organic carbon and available nutrients. Green Gram straw incorporated could easily substitute 50 per cent NPK needs of rice amounting to 60 kg of N, 30 kg of P_2O_5 and 15 kg of K_2O/ha in rice-wheat system without any adverse effect on total productivity.

Although green manuring had been used in India to benefit the succeeding cereal crops, introduction of green manures in rice-wheat systems has been attempted only in the last decade. Green manuring along with recommended level of fertiliser application to both rice and wheat had beneficial effect on the yield of both the crops and increased the total grain productivity of the system by 18 per cent at Ludhiana, 10 per cent at Pantnagar, 6 per cent at Varanasi, 8 per cent at Kanpur and 4 per cent at Bhubaneshwar (Hegde, D.M., 1992). Integrated Nutrient Supply System (INSS) involves the use of chemical fertilisers in conjunction with organic manures/wastes coupled with inputs through biological processes (Singh, G.B., et al, 1990).

Solar Energy

All lines on earth depend on the food generated by green plants. By combining water (from the ground) and carbon (farm the atmosphere) and converting light from the sun into chemical energy, green plants eventually supply all life on earth with the energy needed for their structures and the energy to carry on their life process. Among the various forms of available non-conventional energy sources, solar energy seems to be more promising because of its availability nearly throughout the year in almost all parts of the country.

According to White (1976), the first input of energy in farming takes place during photosynthesis, using the sun's rays at an efficiency of approximately one percent. He used the term 'support energy' and 'energy subsidy' to describe the energy input in crop production which included all items involving an expenditure of energy.

Spedding and Walsingham (1976) used the term 'support energy' to cover 'fossil' fuels and all other sources of energy not directly derived from solar radiation. It also included the energy not directly used in agricultural practices but used in manufacturing agricultural inputs.

Kinne and Macclure (1977) reported that in the primitive agricultural system the process of energy transformation was carried out virtually in their normal state, i.e., the principal energy source was solar, which was augmented slightly by human and animal energy inputs. In modern agricultural system to achieve greater levels of output other energy inputs are used to enhance solar energy transformation.

Indian sub-continent, which stretches from 8 to 36° N latitude and from 6 to 95° E longitudes, has a surface area of nearly 12 per cent of Asian surface area. India, being a tropical country has a significant level of solar isolation over wide regions.

There are about 250 -300 days of useful sub-shine per year in the most parts of the country except in a few regions (Moni A. et al 1982). Considering mean daily global radiation for a year over India, it varies from 5.4 to 5.8 kWh/m^2/day. Moreover vast landscape and geographical location of India offers a good potential for the utilisation of solar energy to ease the stress over conventional fuels.

The sum radiates energy in the form of electromagnetic radiation. Each year an estimated 74.5 ´ 10^{15} kWh of solar energy reach the outer atmosphere, equivalent to about 1.36 kWh per square metre at 160 kilometres above the surface of the earth (Androskey, A. 1973). These figures, however, represent only about one half of one thousand millionth of the total energy radiated by the sun (Bacher, Ken; 1974).

Each year plant photosynthesis fixes about 2 ´ 10^{11} tonnes of carbon with in energy content of 3 ´ 10^{11}J. This is about ten times of the world's annual energy use and two hundred times of our food-energy consumption. In spite of this imposing figure, the photosynthetic process is operating at only 0.1 per cent efficiency of the total incoming annual radiation on the Earth's surface (Gnanam, A; 1990).

For many species, more than 95 per cent of the plant dry weight is derived from photosynthetically fixed CO_2 (Zelitech, 1975). Hence the biomass production is a function of photosynthetic performance and it would be highly advantageous to select photosynthetically efficient plant species for obtaining greater yield. But there are different opinions about photosynthesis and yield relationship among scientists. Several reviews (Good and Bell, 1980; Zelitech, 1982; Nelson, 1988) have discussed various aspects of the photosynthesis and yield relationships (Paliwel et al, 1990).

Plants use radiation between 400-700 nm of the so-called Photosynthetically Active Radiation (PAR). This PAR comprises about 50 per cent of the total sunlight, which on the earth's surface has an intensity of about 800-1000 W/m^2 (David O.Hall, 1980).

Only about 43 per cent of the solar energy reaching the plants in photosynthetically active radiation (of a wave length of 400-700 nanometers) directly involved in reduction of carbon dioxide (Bassham, 1977), and a considerable portion of that (the middle portion) is less active than the wave lengths at either end. For the energy that is successfully captured, nearly one half is consumed by the corn plants in autotropic respiration (Odum, 1971).

This productivity results in an annually energy efficiency for the use of sunlight full spectrum equal to 0.355 per cent. If only the radiation reach the ground during the life of the corn plants is taken into account, the efficiency rises to 0.613 per cent, and if only the photosynthetically active portion of the sun's spectrum is taken into account for the growing season of the corn, the efficiency is 1.426 per cent (Basshan, J.A., 1984).

When solar radiation falls on a well-developed rice crop, about 20 per cent is reflected into sky (Kishida, 1973; Murata et al, 1968). Of the reflected radiation, only about 10 per cent is PAR the rest is in the near far-red region. Thus, 3-5 per cent of PAR is lost by reflection at the canopy surface (Kanda, 1975). Taking 5 per cent as the loss attributed to reflection, about 90 per cent of PAR is absorbed by the leaves and about 5 per cent is transmitted to the ground (Venkateswarulu et al., 1987).

Solar radiation is one of the main consideration in our process energy analysis, because, this is the source of renewable terrestrial energies and the prime mover of photosynthesis (Smil et al., 1983).

The efficiency of solar energy conversion by crops usually approximates 0.2 -0.8 per cent, while in artificial bio-reactors, much higher yields (2-3 per cent) were obtained with some algal cells (Muthuchelian et al., 1988). Research is being conducted throughout the world in an attempt to improve the photosynthetic efficiency of plants. For example, laboratory experiments with algae have succeeded in increasing yields twenty fold over those obtained in ordinary agriculture (Daniels, 1964).

Solar energy consumption or its contribution to the agricultural practices is unviable. The study conducted in this area have very limited focus since the farmers directly use solar energy, only for drying the crop yield and residues. Hence, these energy units were not taken into consideration in the study.

Eco-Friendly Energy

Agriculture is both producer and consumer of energy. Agro-ecosystem receives the energy inputs from the sun, as well as from human sources. This human based energy, however does not feed the biological flow of energy; instead it is utilised in shaping the ecosystem structure. This objective, achieved through decisions on primary productivity, the portion of this production is channelled for human use (Norman, 1978; Shcalezenski, 1984).

For ecological sustainable development, there is a need for developing alternative sources of energy such as bio, solar and wind, which have been neglected all these days because of massive development of the fossil fuel energy. It is heartening to note that the third world countries like China and India have provided the lead in this area by testing and developing biogas plants, mostly in rural areas. In view of the widening gap between the demand and supply of energy and the consequent increase in cost, the use of biogas as alternative source of energy not only supplements energy production but also helps in conserving organic manure for agricultural development (Naidu, K.M.S. et al., 1983).

Human energy and animal drought power contribute as such as 50 per cent of the total power used for traction in agriculture in the third world. The abundant labour supplies of the developing countries also make it more economical than in the developed countries to collect crop and animal residues for use as organic residues for use as organic fertiliser and as commercial or non-commercial fuel, which is more eco-friendly compared with machineries (Mahijani, A., 1975).

Further, these countries should also think in terms of developing other alternative sources, which are indigenously available. There are many areas where the use of animal traction is capable of further development and renewed attention is needed to the improvement of animal draught equipment, which has tended to be neglected in recent years (M.S. Naidu et al., 1983).

2.7 Impact of Energy Scarcity and Rising Prices on Agriculture

Carter and Youde, (1974) explained that the domestic demand for agricultural products would be affected by increased energy prices to the extent that these prices affected economic growth rates, general price levels, and income distribution.

Moriak, (1976) found that in response to the economic stress of increasing energy related costs farmers reduced the use of fertiliser, irrigation, pesticides and herbicides which sharply reduced the yields. It reduced the demand for purchased farm inputs and caused a softening in the price rise of energy related inputs. The economic pressures provided incentives for farmers to improve management practices such that these inputs were most efficiently used.

Warnken, (1976) found that increase in the fossil fuel based agricultural input prices had shifted the comparative advantage of energy intensive agriculture to traditional agriculture in Nicaregua. He observed that a decline in output price would make energy intensive agriculture uneconomical and reduce aggregate output of major food crops, posing series of policy issues for national development.

Mapp and Dobbins, (1976) reported that a rise in natural gas prices increased the cost of pumping irrigation water, other things being equal it reduced the level of net return associated with crop production.

Crosson, (1976) observed that the several-fold increase in the economic cost of fossil fuel spurred the interest in the development of alternative technology in dryland agriculture.

Conor, (1977) found that energy shortages and increased energy prices induced the farmers to reduce the level of energy use in the farm. There were changes in enterprise combinations in response to changing energy prices and supplies. The farmers also shifted to alternative technologies in order to conserve specific types of energy.

Spedding and Walsingham (1978), argued that the future shape of agriculture is common with all other activities would be determined more by the increasing cost of energy than by any other single factor because the price of a resource which was both scarce and essential was bound to have a major influence.

Uri, (1979) estimated the price elasticity of demand for electricity in U.S. agriculture and found that the demand for electricity was relatively inelastic and a large change in price was needed to induce a significant change in quantity demanded.

Simantov, (1980) observed that the rise in the cost of energy affected agriculture as a user and as a producer of energy with different impact in the various regions of a country. A rise in the cost of energy affected in a differentiated way in each line of production and farming system to the extent that within a given farming system, the substitutability of one line of production was made by another. The rising cost of energy had a different impact on the various regions of a country, the region reining large energy costs was penalised in comparison to others.

Beierlein, et al., (1981) found that the demand for electricity and natural gas was price inelastic in the short run and elastic in the long run in all sectors except commercial sector in which the demand was not responsive to its own price.

Yamagida and Conway, (1981) identified two types of energy price effects on agriculture. One was direct price increase pertaining to increase in energy based farm inputs and the second was indirect price increase due to increase in general price level.

Castle (1981), found that in the short run production was made worse off by increased energy cost because, farmers had limited ability to pass on the energy price increase to the consumer, whereas in the long run the increased energy cost could be passed on to the consumer because, prospective food prices have to be high enough to attract and hold resources in the food system. In most cases the system probably could adjust better to higher energy prices than to absolute shortages of energy. Hence, disruptions in energy supply posed a more serious threat to the performance of the food system rather than high prices, but rapidly rising energy prices might sharply reduce the income of some producers within the food system for a period.

According to Lopex (1982), land rental prices might increase in the long run as a consequence of higher energy price. Energy demand for agriculture would decrease if land and energy were substitutable inputs. Chan and Mountain (1983), studied the productivity movements in Canadian agriculture in terms of output prices, energy prices and other material prices. They concluded that an increase in energy price caused a reduction in average labour productivity through increased capital deepening and rising output prices offset the price increase for energy and other materials.

Bhide et al., (1984) argued that an increase in energy price would lead to substitution of land for energy. There was reduction in the rates of fertiliser application and irrigation resulting in reduced crop yields requiring greater use of land to produce a given level of output. Similarly under market condition an increase in energy price would result in higher output price and reduced demand for farm products. If overall demand for the farm product got reduced only less and other farm inputs would be required.

2.8 Energy Conservation

Energy savings are necessary for Indian agricultural farms. Thus, low energy production systems are not well accepted by the farmers who are interested in profitability rather than in energy productivity. A combination of economic and energy analysis of the production system can be more comprehensive for developing the best management strategy.

The new production technologies require a relatively high quantity of inputs, such as fertilisers, irrigation, diesel, plant protection chemicals and electricity. A large number of researchers estimated that energy needs of crops and the extent of renewable and non-renewable energy components of the total energy, particularly in the developing nations are higher (Mittal et al., 1985; Rao, 1986; Ida et al., 1984; Pachauri, 1985; Sinha, 1986). Agriculture in the developing countries in the process of substantial change, there is more scope for the introduction of energy efficient technologies (FAO, 1989).

On Indian farms, energy is available from human, draught animals, electric motors and diesel engines, power tiller, tractors, etc. Energy availability from these sources in different states in 1981 varied from 0.2 to 1.06 kWh/ha with an all India average of 0.4 kWh/ha. The contribution of animate sources of energy was 78 per cent (28 per cent from 146.8 million workers and 58 per cent from 84.6 million draught animals). For crop production (barring irrigation and threshing and shelling operations where electricity and diesel are used to pump water) the main sources of energy are, in fact, animals, humans and to smaller extent, tractors and power tillers (Dhawan and Mittal, 1990).

Some evidences show that weeds affect largely the agricultural productivity in rainfed areas. Estimates show that weeds reduced the productivity of wheat by 15-30 per cent, and in rice by 30-35 per cent, etc. Weeds also remove 30-40, 10-15 and 20-40 kg/ha of nitrogen, phosphorus and potassium respectively (Narayana Moorthy, 1992).

A study by Surrendra Singh (1990) revealed that consumption of commercial energy input has further increased as the area under paddy increased from 1.172 million hectares in 1979-80 to 1.644 million hectares in 1984-85.

And, some studies have analysed the agricultural energy consumption pattern and the types of energy used at macro level (Jones 1988, Veena 1988, Dhawan 1979, Hall and Moss 1983, Smil 1981) and micro level (Reddy 1981, Gupta and Rao 1981, Singh and Banerjee 1988).

Some studies suggested by Tsatsarelis 1993; Gupta, 1984; Smit et al., 1983 and Hopper 1984) have attempted to estimate the energy requirements for each crop. Some others have tried to find out the energy requirements for dry and wet farming (Dhawan and Mittal, 1990 and Rajendran et al., 1987), traditional and modern types of agricultural activities (Coughenou et al., 1985, and Dhanapal, 1991). And then few studies had also tried to find out the energy requirements of different farm size levels (Raikhy et al., 1990), and various mechanisation level (Sahadev Singh et al., 1992; Singh et al., 1975).

Some of the studies have attempted to frame energy policies, chalk out energy plans and suggest optional energy mix. Especially, studies at micro level are very much needed, because energy requirements for agricultural activities differ from area to area depending upon the environmental conditions and agricultural practices. Therefore, many area-based studies are necessary to arrive at meaningful conclusions about energy use pattern in agriculture. Hence, the present study attempts to analyse the energy use pattern in dryland agriculture in a typical block of Tamil Nadu, India.

Moreover, it estimates the energy requirements for different farm sizes with various activities such as ploughing, harrowing, manuring, bunting, seeding or transplanting, fertiliser application, irrigating, weeding, pesticide application, harvesting, threshing, drying, processing and haystack making and determines the influences of energy inputs on output of selected crops of both irrigated and dry farming in the selected villages.

For generating answers to multifarious energy management problems in the farm operations, there are many techniques or tools or models and aids available to a farm energy planner. Some of them are production function, linear programming and farm energy budgeting. The tool can be used under the perfect knowledge and situations. In the present work it is programmed to study the energy analysis for various crops under different farming categories and find out the energy productivity, intensity and efficiency of the energy used for the crops.

Summary

Though many studies on agricultural energy use have been conducted at national and international levels, most of them are concentrated on wet land energy use in different regions and only few studies are in rainfed regions. There were very few studies conducted to compare two regions namely irrigated and dryland cultivation in 'Semi Arid Tropical (SAT.)' areas. In general, the surveys conducted so far differ considerably in the methodology and in terms of information on the mix of fuel, fertiliser and pesticides and choice of mechanisation in the study area. So, it is evident from the literature available that there is a necessity to analyse energy use pattern in different farming system such as wet and dry; farmers category like marginal, small, medium and large; type of operation viz, tractor operated and bullock operated farms. In order to fill up the gap, the present study has been attempted to study the efficiency of energy use in dryland agriculture in Madurai district of Tamil Nadu.

3

Methods and Classification of Energy

3.1 Introduction

The term 'agriculture' includes crop cultivation, livestock rearing and other related farm activities. However, in this study, the crop cultivation alone is undertaken. So, the term 'agriculture' throughout this study means only crop cultivation especially for dryland area.

3.2 Research Design

In this section, at first, the procedure adopted for the selection of the sample is explained followed by methods of energy equivalence and the definition of concepts used in this study. Then, the tools, with the help of which the objectives of the present study are analysed, are presented. Finally, the methods of measurement of variables are discussed.

3.3 Selection of the District and Block

Madurai District is one of the southern districts of the Tamil Nadu. Due to its varied periodic agro-climatic conditions, it possesses the capacity to grow different crops. Hence, Madurai District is purposively selected for the study. Among the 13 blocks in Madurai District, Thevankurichi (T) Kalluppatty block was chosen to prosecute the study due to the following reasons:

(a) Multiplicity of wide range of crops are being grown in this block,

(b) This block covers a sizeable area under wet and rainfed crops,

(c) Primary occupation of majority of the people is agriculture,

(d) The area is more rural based and rainfed.

From this block, 12 village panchayats were randomly selected for the study covering two seasons and by crop concentration identification method (Table 3.1). By this method the researcher can determine the concentration of various crop resources in a given area with the available data. Here the study area map had been chosen and the cultivated areas concentrated by various crops are found by means of location quotient. For example identification of cotton concentrated areas were identified through this method as follows:

$$LQ = \frac{X / Y}{Z / P}$$

LQ = Location Quotient

X = Crop cotton in village - A

Y = All crops in village - A

Z = Crop cotton in Kalluppatty Block

P = All crops in Kalluppatty Block

In the same way the researcher can find the concentration of other crops. The village panchayats selected for this study are Pulkattai, Vannivelampatti, Kunnathur, Velampur, Ganjampatti, Nallamaram, Silarpatti, Sandhaiyur, Melappatti, Pappaiyapuram, Koovalapuram, and Chitulotti.

The farms were selected on stratified random basis. Daily direct and indirect energy inputs into selected crop production activities of 240 crop plots of 240 farmers with farm categories of 12 villages of T. Kalluppatti block in Madurai District and yield of these were entered on a database program for organisation, classification and preliminary analysis of data.

3.4 Sampling Procedure

The study relies on primary and secondary data. Two types of schedules were used for the collection of the necessary data at the primary and secondary levels. One is the village schedule, which was used to collect all information from the village and block officials of the study area, and the other is the farmer's farmhold schedule. The village schedule was distributed to the respective village authorities and with their help the village level details regarding economic and geographic background such as occupational pattern, cropping pattern, irrigation facilities, educational levels, landholding (marginal, small, medium and large farmers) total land available for cultivation, various types of water lifting devices (electrical, diesel and draw wells), livestock particulars, types of machinery or implements available for agricultural purposes and transport facilities were obtained.

The primary data at the farm level were collected from a sample of 240 farms drawn from 12 villages of T. Kallupatti block through pre-tested interview schedule. Apart from energy consumption in the farms of the farmers, size of landholdings, annual income, family size, economic and expenditure pattern and other relevant details were also collected. The interview schedule is designed with due caution to minimise response

Table 3.1

Sample Villages and concentrated crops by Index of Concentrated identification method

Sl. No	*Name of the Village*	*Crops*		*Plots Selected*
		Irrigated	*Unirrigated*	
1.	Pulkattai	Paddy, Cotton, Groundnut	Cotton, Sorghum	20
2.	Vannivelampatti	Paddy, Sugarcane	Cotton, Sorghum	20
3.	Kunnathur	Bajra, Paddy	Cotton, Sorghum	20
4.	Velampur	Paddy, Sugarcane	Sorghum, Cotton	20
5.	Genjampatti	Paddy, Sorghum	Bajra, Pulses	20
6.	Nallamaram	Cotton, Sorghum	Groundnut, Pulses	20
7.	Silarpatti	Paddy, Cotton	Bajra, Sorghum	20
8.	Shandaiyur	Cotton, Bajra	Sorghum, Bajra	20
9.	Melapatti	Paddy, Sugarcane, Cotton	Cotton, Bajra, Groundnut	20
10.	Pappaiyapuram	Paddy, Sorghum	Cotton, Pulses, Bajra	20
11.	Kuvalapuram	Paddy, Sugarcane	Cotton, Bajra, Pulses	20
12.	Chittilotti	Paddy, Cotton	Pulses, Sorghum	20
		Total		240

Source: Compiled based on field study.

error and recall bias. To estimate the energy productivity (kg/kWh), energy intensity (kWh/kg) and energy efficiency (energy output/energy input), the SPSS package was used.

Apart from questionnaire and measurement surveys in all the sample villages, many informal discussions were held with rural households depending on need. During discussions, the various farmers those provide useful information for this study was encouraged to put forth their views freely. In this study PRA (Participatory Rural Appraisal) technique in the respective villages of the study area were used.

The secondary data, regarding the number of implements and machinery distributed, area irrigated by tank and canals, improved seeds distributed for wet and dryland crops, bio-fertilisers (Biological nitrogen fixing algae) i.e., Azosphirillam, Blue Green Algae, Azota, etc., distributed by agricultural extension department were also collected from the records of the Panchayat Union and various related departments of the government of Tamil Nadu. The farmers were grouped into four categories viz., marginal, small, medium, and large farmers, according to their landholding. The various inputs used for each crop were recommended by Agricultural Research Institute, Madurai and Tamil Nadu Agricultural University (TNAU) Coimbatore (Crop Production Guide, 1991). In this context, different authors have used different classification to suit their studies. This classification is done for the present study after analysing the land holding pattern of the study area and given as follows:

Farm Category	*Operation Area (in Acres)*	
	Irrigated	*Unirrigated*
Marginal	< 1	< 2
Small	1 – 2	2 – 4
Medium	2 – 4	4 – 8
Large	> 4	> 8

Among 240 sample farms under various categories, the power available to each farmer was calculated separately on the basis of power sources available to him. Power available per farm was used in place of power available per acre, because the power availability to the farmers usually varies in proportion to the farm holding size. In such a situation a farmer using only animal power may rank higher on a power per acre scale due to very small farm holding size and his mechanisation level on the basis of power available per acre will be misleading.

To determine the energy consumption level in agriculture, the following agricultural activities are included: (1) Ploughing, (2) harrowing, (3) manuring, (4) bunting or seed/farm bed preparation, (5) seeding or sowing or transplanting, (6) fertiliser application, (7) irrigation, (8) weeding, (9) pesticide application or plant protection, (10) harvesting, (11) threshing, (12) drying, (13) processing, (14) packing, (15) haystack making and (16) transport. Further, to measure the energy consumption level in agriculture, the following inputs have been included by converting them into energy units: (a) Human energy, (b) animal energy (c) mechanical energy (tractor, power tiller, thresher, sprayer and other stationary engines), (d) seed input, (e) FYM, (f) diesel, petrol, pesticide/insecticide, electricity etc as shown in the appendix – I.

As discussed above the pre-harvesting energy input was defined as the one applied before the crop matures. The post harvest energy inputs include energy used for harvesting, threshing, crushing, drying, processing, packing, and haystack making and transporting of yield and crop residues. The amount of these inputs was directly proportioned to the quantity of crop handled. The crop yield was assumed to be dependent on the inputs applied before the crop is harvested. The data on energy for transportation are rough approximations, since they are of changing nature ever. Transportation accounts for only a small

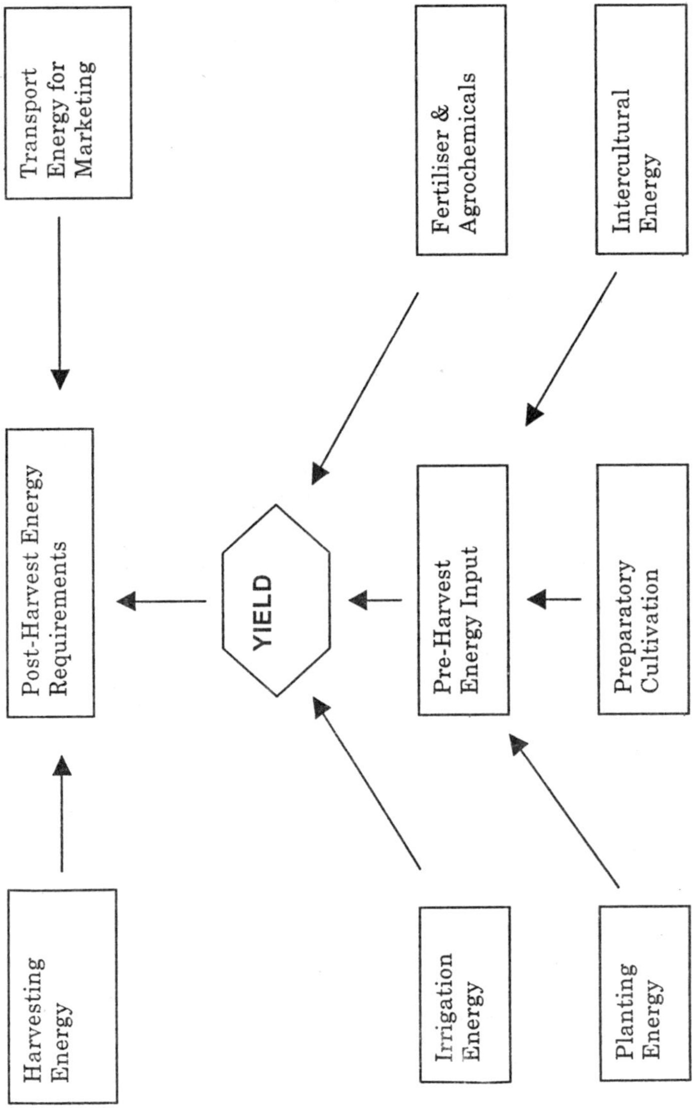
Transport Energy for Marketing
Fertiliser & Agrochemicals
Intercultural Energy
Post-Harvest Energy Requirements
YIELD
Pre-Harvest Energy Input
Preparatory Cultivation
Harvesting Energy
Irrigation Energy
Planting Energy

Figure 3.1

fraction of the total energy consumption, so that this does not affect the analysis very seriously.

Computations of energy inputs and crop yield were done on per acre basis for each individual plot. The use of solar energy for grain drying and other material inputs on plot under study was negligible and since they represent nearly identical geographical locations, the effect of uncontrollable inputs like weather and soil type was assumed to be same on all plots. Inputs and outputs of green gram and black gram were also calculated. These consumed more or less same inputs and produced outputs with negligible difference, so that energy accounting both grams comes under pulses.

Energy inputs were classified according to cultural operations and their functional relation was conceptualised by Sahadev Singh and Gajendra Singh (1972) as shown in the figure 3.1:

3.5 Concepts

The concepts related to energy definition have been dealt by different authors to suit their studies. The present study also have concepts and definitions to support analysis and clear cut understanding are given below:

3.5.1 Energy

A variety of definitions of energy as per their requirements have been used by the studies already carried out in this field. The term energy has been defined in these ways viz., (1) energy to include fuels alone (Sharma and Bhatia, 1989); (2) Energy to include fuels as well as human and animal labour (Reddy et.al (1982), Maheswari et al (1986); (3) Energy includes fuel, fodder and fertiliser in addition to fuels (Bhargavan and Giriappa, 1987). In the present context, the term energy takes into consideration not only fuels but also human energy, animal energy, mechanical energy, wood implements, chemical energy and bio-energy.

Primary Energy: Primary Energy occurs in its natural form before conversion, transmission, and distribution to end use forms. All the energy consumed includes losses at various stages of energy harvesting and processing (Schipper, L 1994).

Secondary Energy: By conversion process, energy can be produced from agricultural by-products and other tree crops. For example mechanical or electrical energy can be generated with the help of biomass gasifiers.

Useful Energy: By useful energy, Schipper (1994) means the difference between the final energy and the estimated conversion losses at the site of final use. Reddy et al (1990) considers useful energy the product of efficiency of end-use device energy input. In this study, the useful energy concept had not been followed for solar energy use. Solar energy actually utilised by crops is calculated after examining the photosynthetic efficiency of the selected crops. Since various equipment like tractor, power tiller, sprayer and other stationary engines, with different efficiencies were used for agriculture purpose; the study discusses gross energy only. It is to be noted here that wherever useful energy is to be referred, it is mentioned as useful energy. So throughout the study unless or otherwise mentioned, energy usually denotes gross energy.

Commercial Energy: Generally it is the energy, which is purchased on a cash-payment basis, and no component of it is produced on the farm. It includes hired human and animal labour power, electricity, diesel, fertiliser, pesticide and other agro-chemicals, machinery, etc.

Non-commercial Energy: It is the energy, which is readily available to the farmers and includes own human and animal labour, farmyard manure, home grown seed, etc. Generally farmers do not pay cash for it. However, in a monetary economy, every farm of energy has its cash value and may be purchased as well as sold in the village itself.

Gross Energy: According to Somasekara (1985), gross energy refers to equivalent energy. Gerald (1985) defines gross energy as the calorific value of the fuel used. Sharma and Bhatia (1987), considered gross energy as the energy content of fuels. In the present study, gross energy refers to the energy content of fuels, i.e., every input is multiplied by its energy equivalents and thus the gross energy is aimed at. The energy content is measured on the basis of the net heating value (also called as the lower heating value) of fuels and for some inputs energy content was taken from standard previous references and expressed in kilo Watt hour (kWh).

Embodied Energy: It means, energy used to produce or manufacture, market and install equipment as contrasted with the energy consumed in putting the goods to use. If we include the embodied energy, the situation becomes even more complicated when evaluating the energy content of non-fuel inputs such as animal draught (Stout, 1989). There is no basis for claiming that a particular choice is right or wrong as long as double counting is avoided; it would be argued that the chosen boundary is the most relevant one for the issue is being studied. The primary obligation of a researcher is to be explicit about the items which have been included and which have been omitted.

For this study, embodied energy was included only for commercial energy such as machinery, electricity, agro-chemicals and fertilisers. The energy embodied in materials was calculated on the basis of industry average values, standard reference values or general usage only.

3.5.2 Classification of Fuels

Energy resources have been classified into various sets such as animate and inanimate; Commercial and non-commercial; traditional and modern; bio-fuels and fossil fuels; conventional and non-conventional; and renewable and non-renewable energy resources by various authors.

In this study, two broad energy classification have been used viz., 1) Commercial energy, which includes diesel, petrol, electricity fertilisers, pesticides, herbicides and machinery 2) Non-commercial energy, otherwise called as, bio-energy, which includes human and animal energy, seed, farm yard manure, green leaf manure, oil cake etc.,

It was further classified on the basis of source as direct and indirect energy.

(A) Direct Source of Energy

The direct source of energy are those which release energy directly like human labour, bullocks, stationery and mobile mechanical or electric motor, power tiller and tractors. The direct energy is further classified as renewable and non-renewable sources of energy, depending upon their replenishment.

Renewable Direct Sources of Energy

In this category, the energy sources which are direct in nature but can be subsequently replenished are grouped. The energetics which may fall in this group are human beings, animals, solar and wind energy and agricultural wastes.

Non-renewable Direct Sources of Energy

In this category, the direct energy sources which are not renewable (at least in near future say nearly 100 years) are included. Coal and petroleum exemplify non-renewable direct sources of energy.

(B) Indirect Sources of Energy

The indirect sources of energy are those which do not release energy directly but do so by conversion process. Some energy is consumed in producing indirect sources of energy. Seeds, manures, chemicals, fertilisers and machinery are classified under indirect sources of energy.

Renewable Indirect Sources of Energy

Seeds and manure were termed as renewable indirect sources of energy as they could be replenished in due course of time.

Non-renewable Indirect Source of Energy

The energy sources, which can not be replenished, come under non-renewable indirect sources of energy such as chemicals, fertilisers and machinery manufacturing.

3.5.3 Agriculture

It is a science of cultivating soil so as to produce economic crops, which demands greater knowledge and skill in its scientific, commercial and artistic aspects. Very broadly, the term is applicable to include both pastoral and arable farming, even though the recent trend is somewhat towards the latter (Gurdeep, 1987).

Traditional Agriculture: Subsistence oriented farming system using low levels of locally available inputs is called traditional agriculture.

Modern Agriculture: Agriculture, which aims at production maximisation through use of external, inputs such as: fertilisers, pesticides, herbicides, mechanisation etc., is known as modern agriculture.

Ecological Agriculture: By ecological agriculture we mean the agriculture that seeks to optimise the use of local farming systems, aiming at a stable, growing and long lasting or non-polluting production level (De Yager, 1992).

Low External Input: An economic approach stressing the need for many farmers to use of techniques that do not require expensive investments from outside the farm is known as low external input.

Farm Yard Manure: Partially decomposed farm wastes to enrich physio-chemical properties of the soil.

Green Manuring: Leguminous plants are grown on field and incorporated into enrich soil fertility (especially nitrogen). This type of manuring is known as green manuring.

Compost: It is a way of decomposing farm and animal wastes for increasing nutrient supply ability of the materials.

Draught Animal: Draught animal is one, which is used to pull carts, trailers, and farm implements.

Pesticides: Pesticides are the energy intensive agricultural inputs. They represent only a small part of the total commercial energy used in agriculture (Mahajan V.S., 1988), since they are mostly used in small quantities.

Bio-mass: Bio-mass refers to all organic matters except fossil fuels. Fuels from bio-mass sometimes called bio-fuels differ from fossil fuels in that they can be rapidly replaced by continued plant production. This energy refers to the energy released when organic matter is eaten, decayed or burnt (Mahajan V.S., 1988).

Bio-gas: Bio-gas is the term used to describe the mixture of gasses produced during anaerobic digestion. The mixture mainly consists of methane (CH_4) and carbon dioxide whose ratio ranges from 1:1 to 2:1. The term bio-gas has been frequently used in the text and it refers to combustible gas containing 50 per cent or more of methane (Chawla, O.P., 1986).

Digested Slurry: It is the residual affluent after fermentation or digestion. It contains some undigested solids and stabilised solids, it is generally darker in colour as compared to the organisational material before digestion and contains all the major and minor plant nutrients. Its addition to the soil improves the biological, physical and chemical properties (Chawla, O.P., 1986)

Bio-fertiliser: Use of micro-organisms to fix or solubilise atmospheric and naturally occurring plant nutrients are called bio-fertilisers.

3.5 Cropping Pattern

It is the rotation of crops taken up in given piece of land, or order in which the crops are cultivated on a piece of land over a fixed period. This is associated with soil management practices such as tillage, manuring and irrigation; and it is normally based on the interest of the farmers and his rights on land extend of land nature of different soils and their extent climatic conditions, agronomic characteristics of crops, availability of labour and power, farmer's own resources, food habits and requirements and marketing conditions. Three types are practised in India: mono-culture or single cropping, mixed cropping and crop rotation (Kamath, M.G., 1972).

Kharif and Rabi Season: There are two main crop seasons in India. Kharif season is a cropping season in which the crops are sown in October or November and harvested in April or May. Sowing in Rabi season begins in the middle of July or August and it is harvested in October or November. This definition is exclusively for Tamil Nadu.

Net Area Sown: Net sown area represents the actual extent of land sown with crops during a given year; when cropped more than once in a year, its area is added to the net sown to arrive at the total area sown during the year.

Principal and Subsidiary Occupations: The principal occupation of sample household is determined by the source of income that occupation, which contributes to more than 50 per cent of the total income of the household, is considered the principal occupation of the household and the other as subsidiary occupation.

3.5.6 Sustainable Development

Sustainable development is defined as 'improving the quality of human life while living within the carrying capacity of supporting eco-system' (IUCN/UNEP/WWP: 1991). Sustainable development is a development strategy that manages all assets, natural resources and human resources as well as financial and physical assets for increasing long-term wealth and well-being (Repetto R., 1986).

This study, however, deal with the sustainable agriculture system which aims at increasing productivity with proper care of environment and ecology.

Productivity: The term productivity is defined as the ratio of the produce in kg to a kWh of energy input. It is expressed in kg/kWh. Here, the produce includes the usable crop output and its residue.

Intensity: Intensity is the reciprocal of productivity. It is given in the units of kWh/kg of the produce.

Efficiency: Efficiency is the ratio of the output energy in kWh to the total input energy used in kWh per acre of particular crop production.

3.6 Method of Measurement of Various Parameters

The measurement of fuel consumption, operation time and crop yields were done in the field. The methods of measurements of different quantities in the study are given below:

Operation Time: A normal clock was used to record the operation time. The time spent on major breakdowns, if any was excluded from time difference between the beginning and end of the operation. However, the time for minor field adjustments was included as the operation time.

Crop Yield: After threshing and uniform drying of the grain, the cleaned crop was weighed on the platform type weighing balance, to measure the yield.

Human and Animal Energy: The method of measuring energy input of human is that to measure the metabolic energy required carrying out certain farming tasks. On an average, the energy value calculated for human and animal labour for work in crop production is about the same as the energy value for food consumed by the labour (Pimental et al., 1979).

Basal Energy: Basal energy requirements of people depend on body size, age, climate, and other environmental factors (Cullumbine, 1950; Quenouille et al., 1951; Suggs & Splinter, 1961; FAO & WHO, 1974). Body size is the most significant of these factors; in fact, a general formula for basal metabolism has been developed:

$$C = aW^b$$

When C = basal metabolism in kcal / day;

A = coefficient, constant for given age and sex, and around 70;

W = body weight in kg and

B = an exponent generally between 0.7 and 0.75

Using a = 70, b = 0.73, and W = 55 kg for a woman, this formula provides a basal metabolism of 0.9 kcal/min. (3.kJ/min.) for the woman and 1.02 kcal/min. (4.27 kJ/min.) for man.

Drawbar pull and speed can be used to measure the power that an animal can produce. The power available for pulling or developing tractive effort is measured at the location where the implement if attached. If the draft and speed of the work performed is known, the drawbar can be calculated as follows (Geo & McDowell, 1980):

$$\text{Drawbar power (kWh)} = \frac{\text{draft (N)} \times \text{speed (km/h)}}{3600}$$

Irrigation

Electricity consumed in irrigating field crops was estimated as the product of the rated horsepower of the pump, the hours of use per irrigation operation and the frequency of irrigation. One horsepower consumed 0.746 kWh of electricity (Ramesh, 1991). The readings of the energy meter at the beginning and at the end of the irrigation were noted, and consumption was recorded.

The human energy inputs were considered for both manual irrigation and the duration of time invested in monitoring irrigation operation. The charges for electric power may be calculated on the basis of number of kWh (1kWh = 1.0 unit) consumed multiplied by the charge per unit. The cost of 1 unit of electricity is Rs.1.25 but in Tamil Nadu farmers are using electricity on free of cost.

3.6.1 Pre-harvesting Operation

The labour inputs for nursery preparation, sowing, transplanting, field operation and ploughing, weeding, fertilising and plant protection were accounted for in estimating the energy requirements for the pre-harvesting operations.

Fuel Consumption

Amount of fuel used for doing particular field operation in case of tractor or diesel engine was measured by top filling method as explained below:

Stationary Diesel Engine: Before starting an operation, the fuel tank of diesel engine was filled up to the marked level. After completing the operation, the engine was stopped and the measured quantity of fuel was added in the tank to fill up to the marked level. The additional amount of fuel put in the tank was measured to get the fuel consumption.

Tractor: The general level of fuel tank of the tractor gets distributed due to undulations of the ground. Therefore, before starting the operation, the tractor is moved on to a levelled ground and its position was marked on the ground. The fuel tank was filled up to the marked level. After completing the operation, the tractor was again moved to the same-levelled ground and the fuel required to fill up the tank to the same initial level was recorded. The fuel consumed is thus calculated.

Farm Yard Manure: FYM is transported on carts adopted for the purpose. FYM inputs were estimated as cartloads. The FYM quantity in a few such carts was estimated by weighing.

Agro-chemicals: The amounts of various fertilisers and pesticides used were recorded. Human energy inputs in the application of the above inputs were estimated. In case of liquid pesticides, a litre of pesticides weighed 0.9 kg.

3.6.2 Post-Harvesting

Labour inputs for threshing, drying, winnowing and packaging were quantified. Transportation was included in the present study as labour, bullock carts or tractors involved in moving the produce from the fields to the granaries or residences. Moreover, some of the products were transported to areas well outside the defined system boundary.

3.6.3 An Estimation Cost of Cultivation

In this study, the cost of cultivation included material costs and energy costs. Material costs included value of seeds, manures and fertilisers and plant protection chemicals. Energy cost included costs of human labour, bullock labour, electricity and hire charges on machinery. The costs were computed at the prevailing market rates. Gross income was defined as the total value of both main product and by-product valued at harvest prices. Net income was worked out by deducting the cost of cultivation from income.

3.7 Tools of Analysis

The tools used for analysing the data include conventional techniques such as percentages, averages, ratios and multiple regression models. To determine the effects of energy consumption level in agriculture, the regression co-efficients model for yield has been developed with operation-wise and category-wise classification with five independent variables and one dependent variable. The independent variables are (1) area (wet and dry), (2) animate energy (human and animal), (3) Bio-energy, (4) mechanical energy (machinery including wood instrument, oil and electricity), and (5) chemical energy and the dependent variable is yield. Thus, by using SPSS package, stepwise multiple regression analysis is selected since it is really a search procedure; it enters each variable one at a time into the regression equation in the order, the greatest contributor being entered first. Thus, out of any number of independent variables, the stepwise technique "searches out" the greatest contributor to the total variance and effectively rank-orders them.

3.8 Summary

Energy is the basis of economic activity and largely determines the living conditions of a country's inhabitants. Every resource is often considered as national assets and sufficient energy production is a pre-requisite to development. In the developing countries, pattern of energy use and changes in them are very different in wet and dryland agriculture, and vary among regions as well. Changes in energy consumption over time are determined by several factors. The factors that have a major impact on the amount of energy consumed by agriculture sector and fuels used include income, farm size, labour utilisation, price or personal costs of obtaining fuels, availability of modern fertilisers and efficiency of end use mechanical equipment. Most of these factors/concepts are interrelated and major implementations for the policies aimed at solving energy problems of rainfed farming system. This study is programmed to cover all aspects in the dryland crop cultivation.

4

Socio-Economic and Farmholding Practices

4.1 Introduction

This chapter contains two sections. In section one, a brief socio-economic profile of the study area is presented. Important characteristics of the sample landholders are presented in section two. In both the sections, due emphasis has been accorded to the features that have an influence on the pattern of energy consumption.

4.2 Area Profile

4.2.1 Location

The study area, Thevankurichi (T) Kalluppatti block is situated along the east-south direction about 46 KM away from Madurai city, on the national highway to Courtalam. It extends 15°25' North to 15°35' South and 77°40' East to 77°52' West. This block is located in the eastern slope of the Western Ghats, which is named as Sadhuragiri Hills. The adjoining blocks are Sedapatti in the West, Tirumangalam block in north, Kalligudi bolck in the east, Sivakasi and Watrap blocks of Virudhunagar district are in the south boundary.

4.2.2 Landscape

T.Kalluppatti block is a plain area but there are ups and downs in some places. This block extends in the northwest direction up to Chempatti village, which is located at the foothills of the Sadhuragiri and Sengulam in the Southeast. There are two types of the soils are found in the study area namely red loamy and black cotton soil.

4.2.3 Land Classification

Land classification the district, block and the respective villages was assessed from village records and concomitant ground surveys in the study area. Total land available in the district is 374812 hectares and 26532 hectares in the block. Of which, area under net cultivation covers 58.53% at the block level and 46.06% at the district level. Others contribute 53.94% in the district and 41.47% at the block level which includes current fallow, forest, uncultivable land, cultivable waste, land put under non-agricultural purpose and land for miscellaneous tree and other crops.

4.2.4 Climate and Rainfall

Normally the climate, in this block is pleasant during spring season and hot in the summer season. It has usually a favourable agro-climatic condition conducive to a number of food and non-food crops. Accordingly there are three seasons in this block, viz., Kharif, Rabi, and summer.

Kharif: This season begins in June, which is extended up to October. Kharif season is also known as south-west monsoon season. Major crops grown in this season are paddy, sorghum, bajra, groundnut, cotton, sugarcane and senna. Paddy also been cultivated in the study area based on the availability of water in the local tanks/ watersheds.

Rabi: Sowing in Rabi season begins in the middle of October, which extends up to April. This season is otherwise called north-east monsoon, and the major crops grown in this season are senna, millets and all grains.

Summer: The summer season begins in the late March and extends up to June. Cotton, maize, groundnut and senna are grown in summer season also. Generally, the climate is hot and dry. The hottest period is from April to May. The temperature in the block gets reduced during the north-east monsoon period. The average normal rainfall for the block is 827.1mm/year. But in the year 1997-98 the average actual rainfall was 952.9 mm which is comparatively high in the year.

4.3 Demographic Characteristics

The Kalluppatti block is one among 13 blocks of Madurai District. This block comprises of 42 village panchayats and 48 revenue villages with 181 hamlets. The region spreads over an area of 206.15 square kilometres and has a population of 80479 of which 40373 (50.17%) are male and 40106 (49.83%) are female according to the 1991 census. The density of population is 390 persons per square kilometre. This block has started to function from 2nd October 1961. Location of the study area, sample villages, share of irrigation on cultivation and classification of

Table: 4.1
Classification of Village Panchayats by Population Size

Sl. No.	*Village Particulars*	*Population as per 1991 Census*					*Total*
		Below 1000	*1001 to 2000*	*2001 to 3000*	*3001 to 4000*	*Above 4000*	
1.	Number of Village Panchayats	10	22	7	2	1	42
2.	Number of Revenue Village	6	26	10	5	1	48

Source: Block Census Book, 1991.

land area in respect to this study are given in the figures 4.1-4.4.

4.4 Villages Classified by Population Ranges

In the study area, there are 42 village panchayats and 48 revenue villages. According to the recent (1991) census the village panchayats have been classified into four categories based on the population as shown in following table (4.1).

Most of the revenue villages are under 1000 to 2000 population range, and only one village is under the category of above 4000 population.

4.5 Educational Institutes

Education up to college, higher secondary, and high school levels was present in the study area. There are primary schools in twelve villages and middle schools in three villages. In Modagam alone, one primary school and one middle school are functioning. In total, in the twelve sample villages, there are 27 primary schools (up to fifth standard) three middle schools (up to eighth standard) two high (up to tenth standard). There are 3 higher secondary school in the selected block and no collegiate education centre even in the block. Students have to go either to Madurai or Virudhunagar for their collegiate education and their professional and university education, which are the head quarters of the respective district and its neighbour respectively.

4.6 Agriculture

The vast majority of the people in this block are engaged in dryland agriculture. A variety of crops are grown in the block. Paddy is the major food crop occupying considerable area of total wet cropped area of the block. Total geographical area of the region under investigation is 26532 hectare (Table 4.2). The total area under cultivation is 18232 hectare out of which 2702 hectare (14.8%) comes under wet land and 15530 hectare (85.2%) comes under dry land (figure 4.3) is the major evidence for the dryland cultivation in the study area.. Among 42 village panchayats in the study area, majorities of the wet lands are evenly distributed.

4.7 Irrigation

There are two types of irrigation methods followed in the study area viz., surface irrigation and sub-surface irrigation. Surface irrigation is fully used in the monsoon season especially for paddy cultivation. In the 12 villages, there are 16 PWD tanks, 4 minor irrigation tanks and 116 irrigation tanks. Sub-surface irrigation is drawing water from wells or tube wells. It requires equipment for lifting the water from wells. T.Kalluppatti block has 285 wells. Out of which 143 wells are not suitable for irrigation. Of the remaining wells 113 are energised with electric motor, 17 are energised with oil engine and 12 are bullock operated wells.

4.8 Administration

The T. Kalluppatti Panchayat Union is under the administration and supervision of the Block Development Officer (BDO) and he is assisted by the Divisional Development Officer. There is another Block Development Officer under the control of District Rural Development Agency (DRDA), specifically for the purpose of implementing rural development schemes for the block.

There are three assistants and two record clerks, three junior assistants and five office assistants to help the BDO in general administration and supervision. There are seven extension officers who are in-charge of various sectors of Rural Development Programmes. There is one extension officer who is completely in-charge of the industrial sector. Seven rural welfare officers, one assistant and one accountant assist him. One educational officer is looking after the social education and public relation work. He is also in-charge of family welfare and small savings and other social works. The another extension officer, who is in-charge of Adhi Draviders, look after the development programmes of Adhi Draviders. There is also an extension officer, who is in-charge of social forestry programmes. There is a separate extension officer for agriculture and co-operation and he is the in-charge of experimental plot

Table – 4.2
Area Profile - A Glance

Item	*District*	*Block*	*% to the District*
Population (1991 Census)			
Male	706728	40373	5.71
Female	685236	40106	5.85
Total	7752464	80479	1.04
Total Villages	596	181	30.37
Village Panchayats	419	42	10.02
Town Panchayats	15	2	13.33
Geographical Area (ha)	374812	26532	7.08
Net Cultivation	172651	15530	9.00
Current Fallow	45515	4659	10.24
Forest	58930	1051	1.78
Area Under Non-Agri.	59381	3286	5.53
Uncultivable	19470	1028	5.28
Cultivable Waste	5600	106	1.89
Area in Other Use	13265	872	6.57
Net Irrigated Area (ha)	105666	2702	2.56
Canal Irrigation	53716	1030	1.92
Well Irrigation	30921	1479	4.78
Other Irrigation	21022	193	0.92
Major Crops (ha): *Paddy*	88943	1255	1.41
Groundnut	16433	1353	8.23
Bajra	6024	81	1.34
Sugarcane	7813	96	1.23
Cotton	14489	4663	32.18

(Contd.)

(Contd.Table 4.2)

Item		*District*	*Block*	*% to the District*
Size of Landholdings (No.)				
Below 1 ha.		258311	15486	5.99
1 - 2 ha.		39952	3319	8.30
2 - 4 ha.		16404	1111	6.78
4 - 10 ha.		4403	233	5.29
Above 10 ha.		406	14	3.45
Total		319476	20163	6.31
Others	*Total Animals*	965865	14551	1.51
Tractors		609	67	11.00
Commercial Banks		101	6	5.94
Co-operative Banks		222	1	0.45
PACBs		209	29	13.88
PHCs & Hospitals		94	8	8.51
Schools		1188	87	7.32
Industries	*SSI*	680	29	4.26
Cottage / Village		5534	1438	25.98
Rural Markets		6	1	16.67

Source: Compiled from various Sources (1997-98) at the District level.

supervisors. The prime duty is to disseminate all the improved technology from lab to land. Under government subsidy scheme, they are distributing seeds, fertilisers, power sprayers and other agricultural inputs and implements. At the village level, the Village Administrative Officers (VAOs) are the ultimate revenue authority, and are responsible for land revenue collection.

4.9 Number and Area of Operational Holdings

The following table (4.3) shows 10 classification of land sizes in the study area. Among total number of farmers, 53.81 per cent hold that up to 0.50 hectare of land and 30.13 per cent of the farmers hold 1-2 hectare of land. Farmers who possess

Table – 4.3

Number and Area of Operational Holdings (1997-98)

Sl.No	*Size Class (in Ha)*	*Numbers*			*Area in Hectare*		
		SC/ST	*Others*	*Total*	*SC/ST*	*Others*	*Total*
1.	Up to 0.50	530	10320	10850	111.66	2745.46	2857.12
		(71.05)	(53.15)	(53.81)	(29.02)	(18.49)	(18.76)
2.	0.50 - 1.00	125	4511	4636	88.00	3303.83	3391.83
		(16.76)	(23.23)	(22.99)	(22.87)	(22.26)	(22.28)
3.	1.00 - 2.00	77	3242	3319	104.88	4483.42	4588.30
		(10.32)	(16.70)	(16.46)	(27.26)	(30.21)	(30.13)
4.	2.00 - 3.00	6	800	806	14.24	1887.45	1901.69
		(0.80)	(4.12)	(4.00)	(3.70)	(12.72)	(12.49)
5.	3.00 - 4.00	2	303	305	7.84	1049.06	1056.90
		(0.27)	(1.56)	(1.51)	(2.04)	(7.07)	(6.94)
6.	4.00 - 5.00	1	127	128	4.05	564.95	569.00
		(0.13)	(0.65)	(0.63)	(1.05)	(3.81)	(3.74)

(Contd)

(Contd. Table 4.3)

Sl.No	Size Class (in Ha)	Numbers			Area in Hectare		
		SC/ST	Others	Total	SC/ST	Others	Total
7.	5.00 - 7.50	2 (0.27)	77 (0.40)	79 (0.39)	11.58 (3.01)	452.11 (3.05)	463.69 (3.05)
8.	7.50 - 10.00	--	26 (0.13)	26 (0.13)	--	218.00 (1.47)	218.00 (1.43)
9.	10.00 - 20.00	2 (0.27)	10 (0.05)	12 (0.06)	22.03 (5.73)	117.44 (0.79)	139.47 (0.92)
10.	Above 20.00	1 (0.13)	1 (0.01)	2 (0.01)	20.50 (5.33)	20.50 (0.14)	41.00 (0.27)
	Total	746 (100)	19417 (100)	20163 (100)	384.78 (100)	14842.22 (100)	15227 (100)

Source: Block Census Book and Land records of the respective village Panchayats (1997-98).
Note: Total area available for cultivation is 15530 hectares, but the area under different classes of land holding is 15227 hectares. The remaining 303 hectares of land are in the records of the village panchayats and temples.

10-20 and above hectares of land are very less in number i.e., 0.92 and 0.27 per cent respectively. From this table we conclude that most of the farmers come under small and marginal farm category. The marginal farm holders constitute vast majority compared to other categories.

4.10 Area Under Different Crops

The following table (4.4) shows that the area under different crops in the study area for the year 1997-98 and also

Table – 4.4
Area under Different Crops in the Block

Sl.No	*Crops*	*Area Under Cultivation (Ha)*		
		1997-98	*1996-97*	*Difference*
1.	Rice			
	Main Season: Local Variety	479	341	138
	Hybrid Variety	614	577	37
	Second Season:			
	Local Variety	162	--	162
	Hybrid Variety	--	--	--
	Total	1255	918	337
2.	Sorghum			
	First Season:			
	Local Variety	643	469	174
	Hybrid Variety	--	406	-406
	Summer Season:	--	--	--
	Total	643	902	-259
3.	Bajra Main Season:	81	79	2
	Second Season:	--	44	-44
	Total	81	123	-42
4.	Ragi	14	--	14

(Contd.)

(Contd.Table 4.4)

Sl.No	*Crops*	*Area Under Cultivation (Ha)*		
		1997-98	*1996-97*	*Difference*
5.	Pulses	1537	1061	476
6.	Sugarcane	498	495	3
7.	Cotton	1498	1112	386
8.	Groundnut	640	750	-110
9.	Others	2577	780	1797
10.	Total Food Crops	7641	14202	-6561
11.	Total Non-food Crops	9709	4357	5352
12.	Crops Sown more than once	301	1510	-1209
	Net Sown area	15530	14049	1481

Source: Block Census Book and Taluk statistical office, (1997-98).

depicts the difference between the earlier year. Area under cultivation was higher (15530 ha.) in the year 1997-98 while, area under cultivation was somewhat low in 1996-97, the difference being 1481 hectare. This is because of the increased rainfall and other natural factors.

Comparing with other crops, area under cultivation of non-food crops was higher, the difference registered being 5352 hectare. Normally, it requires less water than other food crops like paddy. From this we conclude that the area received good rainfall in the monsoon season during the year 1997-98. In the year 1996-97, there was erratic distribution of rainfall in the different seasons. So area under cultivation of rabi crops were affected. Major crops grown in the season were cotton, senna, pulses and sorghum.

4.11 Livestock

One cannot deny the fact that cattle have been playing a key role in our country since early millenniums or rather through the development of civilisation. But it will not continue to play an important role in the years to come in the study area due to the cattle exports to neighbouring state.

There are 23 milk producers' coop. societies functioning and hence there is good scope for dairy farming. For prevention against diseases, facilities such as veterinary hospitals, dispensaries and veterinary sub-centres are available in the block at T. Kalluppatti and Peraiyur and four artificial insemination sub-centres which are located in Melapatti, Silarpatti, Vannivelampatti and Karaikeni.

The table (4.1) of area at glance shows that the share of total animal population in the block is very low (1.51%) i.e., 14551 heads registering when compared with the district level (965865). It led most of the farmers to forcibly depend on mechanisation of agriculture.

4.12 Sources of Irrigation

In the study area, only 2701.98 hectares of land were irrigated by various sources in the year 1998 whereas 1692.28

Table – 4.5
Sources of Irrigation (in Hectares)

Sl. No	*Sources of Irrigation*	*Area in 1998*		*Area in 1997*		*Difference*	
		Total	*Net*	*Total*	*Net*	*Total*	*Net*
1.	TANK:						
	Government	748.85	680.48	656.69	612.8	92.16	67.68
	Private	--	--	--	--	--	--
	Total	748.85	680.48	656.69	612.8	92.16	67.68
		(25.94)	(25.18)	(26.99)	(36.21)	(20.32)	(6.70)
2.	WELL:						
	Government	--	--	--	--	--	--
	Private	1943.61	1912.89	1663.26	981.96	280.35	930.93
	Total	1943.61	1912.89	1663.26	981.96	280.35	930.93
		(67.34)	(70.80)	(68.36)	(58.03)	(61.81)	(92.20)

(Contd.)

(Contd. Table 4.5)

Sl.No	Sources of Irrigation	Area in 1998 Total	Net	Area in 1997 Total	Net	Difference Total	Net
3.	WELL (For Allied Irrigation):						
	Government	--	--	--	--	--	--
	Private	194.08	108.61	113.04	97.52	81.04	11.09
	Total	194.08	108.61	113.04	97.52	81.04	11.09
		(6.72)	(4.02)	(4.65)	(5.76)	(17.87)	(1.10)
	Total Irrigated Area	2886.54	2701.98	2432.99	1692.28	453.55	1009.7
		(100)	(100)	(100)	(100)	(100)	(100)

Source: Block Census Book and Taluk Statistical Office (1997-98).

hectare was irrigated in the year 1997 (table 4.5). The area under irrigation by different sources increased from 1996-97 to 1997-98 because of considerable rainfall during the year. Net area irrigated by tank was 680.48 in the year 1998, which was more than in 1997. The net area irrigated by different irrigation sources was 58.03 per cent in 1997 and 70.80 per cent in 1998.

Table – 4.6
Agricultural Implements and Machinery

Sl.No.	Items	Numbers
1.	Plough	
	(a) Wooden	1847
	(b) Iron	169
	(c) Total	2016
2.	Water Pump for Irrigation Purpose	
	(a) Worked by Oil Engine	278
	(b) Worked by Electric Motor	3459
	(c) Total	3737
3.	Tractors	
	(a) Government	2
	(b) Private	65
	(c) Total	67

Source: Block Census Book (1996-97) and Agriculture Extension Department.

4.13 Agricultural Machinery and Implements

Agricultural operations have been carried out with the help of various implements. They are the plough, the irrigation pump, the tractor and other stationary equipment or engines. For preparatory cultivation, two types of plough have been used such as wooden and iron plough. The availability of wooden plough is higher (1847) than the iron plough (169) among various categories of farms in different villages (table 4.6). 3459 wells were energised to lift water. Among these 278 had oil engines and remaining 3359 had electric motors. Here electric motors were more predominantly in use to lift the water than through oil engines. For ploughing, threshing and transporting purposes, there were 67 tractors available in the study area. Among these 67 tractors only two was owned by the government i.e., agro-service Centre, farmers owned the 65 tractors. For sugarcane crushing, almost all the farmers are sending their produces to the nearby sugar mills, since no crusher in the study region.

4.14 Characteristics of Sample Farmers' Farms and Households

For the present study 240 farmers are taken as sample, from 12 village panchayats. As seen in the chapter on research design, in all the sample villages, due care was taken to make the sample representatives. In this section, the characteristics of the sample farmers' farms and households are provided.

4.14.1 Age and Education of the Sample Farmers

The farmers (240) are distributed over four age groups covering from 24 years to 55 years and above (table 4.7). Among the marginal farm, the majority of them are in the age group 55 and above (34.78 per cent). The second largest age group with 28.99 per cent is in the 45-54 years range. The age group up to 35 years and 35-44 years form the minority with 11.59 and 24.64 percents respectively.

Table – 4.7

Age and Education of the Sample Farmers

Variable in Years	*Marginal Farmers*		*Small Farmers*		*Medium Farmers*		*Large Farmers*	
	No.	*%*	*No.*	*%*	*No.*	*%*	*No.*	*%*
Age								
Upto 35	8	11.59	4	5.80	13	16.46	0	0.00
35 - 44	17	24.64	21	30.43	19	24.05	4	17.39
45 - 54	20	28.99	19	27.54	24	30.38	12	52.17
55 & above	24	34.78	25	36.23	23	29.11	7	30.43
Total	69	100	69	100	79	100	23	100
Education								
Illiterate	6	8.70	4	5.80	10	12.66	1	4.35
Primary	35	50.72	28	40.58	43	54.43	8	34.78
Secondary	24	34.78	36	52.17	23	29.11	12	52.17
Collegiate	4	5.80	1	1.45	3	3.80	2	8.70
Total	69	100	69	100	79	100	23	100

Source: Primary data.

Among the small farmers, the majority 36.23 per cent falls under the age group of 55 years and above, while the years up to 35 age group forms the minority with 5.80 per cent each. In the same way medium farmers form majority (30.38 per cent) under age group of 45-54 years, whereas the majority (16.46%) is in the age group of below 35 years.

Next, among the large farmers, the majority 52.17 per cent comes under the age group ranging between 45-54 years while the minority 17.39 per cent falls under the age group between 35-44 years. There are none in the age group of up to 35 years among large farmers.

The table - 4.7 shows educational particulars of the sample. The following can be inferred among the marginal farmers, 8.70 per cent are illiterate, 50.72 per cent have received primary education and 34.78 and 5.80 per cent have received secondary and collegiate respectively. Among the small farmers 5.80 per cent are illiterates, 40.58 per cent have received primary education, 52.17 per cent have received secondary education and only one (1.45%) have gone through the collegiate education.

Among medium and large farmers, 12.66 and 4.35 per cent have received primary education respectively. Among large farmers 8.70 per cent have received collegiate education and 4.35 per cent are uneducated. The rate of illiteracy is very less among the large farmers, i.e. just 9.6 per cent.

4.14.2 Characteristics of Land Holdings

The land owned by the medium farmers in the study area is 76.76 per cent, 9.92 per cent of the land owned by small farmers, 8.95 per cent of the land owned by the large farmers and 7.03 per cent of the land owned by the marginal farmers are covered with irrigated farm (table 4.8).

37.94 per cent of the land owned by the small farmers, 28.79 per cent of the land owned by the marginal farmers, 20.59 per cent of the land owned by medium farmers and 12.67 per

Table – 4.8
Landholding Pattern of Sample Farmers

Sl.No	*Category of Farmers*	*Irrigated Farm*		*Dry Farm*	
		No. of Farmers	*Land Size (Acre)*	*No. of Farmers*	*Land Size (Acre)*
1.	Marginal Farmers	4	10.00 (7.03)	65	97.25 (28.79)
2.	Small Farmers	9	14.10 (9.92)	60	128.15 (37.94)
3.	Medium Farmers	42	109.15 (76.76)	37	69.55 (20.59)
4.	Large Farmers	5	8.95 (6.29)	18	42.80 (12.67)
	Total	60	142 (100)	180	337.75 (100)

Table – 4.9
Sample Farmer's Livestock Particulars

Sl. No	Category	*Marginal Farmers*		*Small Farmers*		*Medium Farmers*		*Large Farmers*	
		No.	*%*	*No*	*%*	*No.*	*%*	*No.*	*%*
1.	Milch Animals (Total – 300)	124	41.33	86	28.67	38	12.67	52	17.33
2.	Drought Animals (Total – 395)	89	22.53	219	55.44	71	17.97	16	4.06
3.	Sheep and Goats (Total – 767)	112	14.60	364	47.46	243	31.68	48	6.26
4.	Others (Total – 1261)	516	40.91	490	38.86	135	10.71	120	9.52
	Total (2723)	841	30.88	1159	42.56	487	17.88	236	8.67

Source: Field data.

cent of the land owned by large farmers come under dry land agriculture. It results that the average size of land holding and major ownership of the land covers with medium and small farmers in respect of irrigated and unirrigated farms.

4.14.3 Sample Farmers' Livestock Particulars

The total of milch animals owned by all farmers is 300. Of this 41.33 per cent of the milch cattle is owned by marginal farmers, 28.67 percent by small farmers, 12.67 per cent by medium farmers and remaining 17.33 per cent by large farmers (table 4.9).

Total draught animals owned by sample farmers is 395, out of which, majority of the animal (55.4 per cent) are owned by small farmers, remaining 44.56 pre cent being owned by the other three categories. Total goats and sheep owned by sample farmers are 767, out of which, 47.46 per cent are owned by small farmers, 31.68 per cent by medium farmers, 14.60 per cent by marginal farmers. The large farmers are having only 6.26 per cent goats and sheep.

Animals other than cows, sheep and goats (including poultry and other domesticated animals) owned by the four categories of farmers are 1261. Marginal farmers are having 40.91 per cent while large farmers own 9.52 per cent only.

4.15 Summary

The profile of the study area paves the way for planning region and area specific models to be suit to each region. This necessitates determining priorities on the basis of the short-term and long-term energy needs of the regions, keeping in view the available human, animal, capital and mechanical/ material resources. This is particularly true where it is necessary to check the consistency of targets for the model. It will be of immense help to formulate a proper development policy for the successes of a development of energy plan and to avoid any pitfalls that may arise in the agricultural development.

It is meant to give appropriate suggestions regarding agricultural energy in different regions with different category of farming classes. Thus this chapter is a useful tool which highlights the importance of resources and helps in solving the dryland energy problems.

5

Energy Use Pattern in Dryland Areas

5.1 Introduction

This chapter presents an analysis of the data collected from various categories of farmers based on the sample survey related to the study area. The tables present average results for each crop from the 240 respondents. The study involves the consumption of commercial and non-commercial energy, irrigated and unirrigated, seasons, and different farmers in various categories of farms etc.

In order to achieve good returns from the crop, the cost of cultivation must be kept low. To do this, economic analysis must be carried out to identify, where energy savings could be made without affecting yield or profitability. Naturally this is to be done through energy analysis, which indicates ways in which energy efficiency could be increased. The low energy production systems are not well accepted by the farmers who are interested in monetary benefits rather than energy productivity. A combination of economic and energy analysis of the production system could be a comprehensive way to analyse the best management strategies (Raja R. 1995).

The following sections deals with the crop-wise energy consumption analysis. First, source-wise input energy consumption of various operational and category-wise farm has been analysed. Second, these results are compared with the results from both irrigated and rainfed farming. Third, output energy production has been calculated to both main and by-products of all the seven selected crops irrespective of the dry and wet farms, using appropriate energy equivalence. Fourth, the energy productivity, intensity and efficiency of energies have been critically examined and compared with various farms. Finally, important results are highlighted to all the crops. Here the main results are discussed while the main tables are reported in the attached appendix-II.

5.2 Paddy: (Oriza Sativa)

Rice is the most important cereal crop grown over an area of about 42 million hectares in India with irrigation coverage of only 43 per cent. With the good monsoons of 1990-91 and 1992-93, the country has been able to set up its production considerably with a peak production of 74 million tonnes in 1992-93 (Economic Survey: 1992-93).

Though India is a major rice producer next of China and it has increased its production from 22 million tonnes in 1949-50 to 74 million tonnes in 1990-91. Productivity is very low when compared to many countries like China, Egypt, USA, Italy and Japan (Biswas et al: 1991). Productivity increase through integrated and efficient management of energy inputs, therefore, needs due attention.

In the study area, paddy has been grown under irrigated conditions only. Table 5.1 in the appendix-II shows that the source-wise energy requirements in paddy production per acre among various categories of farms in two types of operation (bullock, and tractor operated farm). A few inputs such as seed, FYM and other chemical energies are used with a little variation by all categories of farmers. But marked variations have been identified in the use of animate energy (human and animal), instruments, diesel and petrol, and machinery in all farms. This table 5.1 clearly indicates (appendix-II) that, bio-energy utilisation is low among all categories of farms. But the utilisation of total commercial energy is more by all categories of farms, than non-commercial energy for all kinds of operations.

5.2.1 Bullock Operated Farm

In this farm total direct and indirect sequestered energy of various categories of farmers are calculated to be 10145.27, 9239.80, 9077.72 and 10388.77 kWh/acre of marginal, small, medium and large farmers respectively. Energy needed for nursery raising is also included in the various sources of total energy. Energy consumption is high in large farmers (LF) followed by Marginal farmers (MF). But Medium farm (MeF) consumed less amount of energy among all bullock operated farms on paddy production. All farms use less commercial energy such as chemical fertilisers, pesticides and machinery compared with Tractor Operated Farm (TOF). On the other hand, it is minimum in the case of bio-energy use in both the farm operations.

Electricity, which is consumed only for irrigation purpose, has an average share of 11.79 per cent in energy consumption in Bullock Operated Farms (BOF), whereas less amount of electrical energy used by various categories, except marginal farms (Table 5.1). In the study area, the Government of Tamilnadu is supplying electricity for free of cost.

In bullock operated farms, non-commercial energy use is higher when compared with other farms i.e. tractor operated. As far as the total energy input is concerned, it is the bullock operated farms, which register the maximum. In this, major portion (one third) contributed by bio-energy, does not affect the ecology of the system. There could be a possibility to increase the bio-energy use with considerable reduction in the external input (commercial sources) without reducing yield.

5.2.2 Tractor Operated Farm

In traditional agriculture man power and animal power were the two major sources of energy for carrying out various agricultural operations. This was more time consuming and cumbersome; but it consumed more of non-commercial energy. As a result of the technological advancements, agricultural machinery has been introduced to replace and in some cases to supplement animal and human labour. This has brought a significant change in the pattern of energy use in agriculture sector. The second part of table 5.1 shows the changes in the technology, which made significant change in the pattern of energy utilisation.

The most important machine, which is used in the modern agriculture, is tractor. It is clear from table 5.1 that machinery energy consumption is obviously more in tractor operated farm than that of bullock operated farm, whereas bio-energy utilisation is high when compared to the other farm. Due to this reason, total energy utilisation is low in tractor operated farms, which affects the agro-ecosystem.

In the large farm, total energy consumption is high (9208.24 kWh/acre) but in the other categories of farmers the same are 9203.99, 8315.56, 8157.37 kWh/acre for medium and large farmers respectively. These are all well below the prescribed level by the agriculture scientists. Generally total energy consumption in a tractor farm was less than that in bullock operated farms; and the commercial energy is also found to be lesser in tractor farms. From this, one can conclude that for paddy cultivated areas require more non-commercial energy.

The following are the inferences drawn from Table 5.1.

(a) Total energy consumption is high in bullock operated farm; out of this total, one third of the energy is from non-commercial sources.

(b) Commercial energy consumption is less in both bullock-operated and tractor operated farms.

(c) In the study area, some of the farms are irrigated by wells, and so, it requires electricity for lifting water, electricity consumption is high for paddy than other crops.

(d) The inputs like seed, FYM, fertiliser, pesticides and electricity are more or less same in all farmer categories.

(e) Use of animate energy is more in bullock operated farms, whereas it is very small in tractor farm. The use of wood instruments is negligible in tractor farms.

5.2.3 Operation-wise Energy Requirements in Paddy

Table 5.2 in the appendix-II indicates various sources of energy distributed in different operations in paddy cultivation. In the two types of farms surveyed, bullock operated farm consume large amount of energy than the other farm. In the case of total energy consumption for preparatory cultivation levels, among these farmers, small farmer consumes 1054.56 kWh/acre, which is lesser than the others; but in bullock operated farm, large farmer category alone exceeds regaraing total energy consumption (10388.77 kWh/acre).

Seed input, irrigation, fertiliser, plant protection and manure application are more or less same in all farms. Apart from this, the remaining operations like preparatory cultivation and harvesting consume more energy in bullock operated farms. This is due to more animate energy used for ploughing in preparatory cultivation and threshing in harvesting. The use of animate energy is a time-consuming process compared to work done by the machineries, but it will not create any effect on the agro-ecosystem.

In tractor operated farm, transportation consumes more energy, but the energy consumption is insignificant for preparatory cultivation. Even though this farm consumes lesser total energy compared to bullock operated farms, the energy comes from commercial sources like petrol and diesel constitutes major portion. The following are the findings from table 5.2.

1. In all the farms, maximum amount of energy is consumed by manuring, which is more than 40 per cent of the total energy.

2. Among various field operations in the paddy production, commercial energy is utilised in less number of operations. The various operations are irrigation (electricity based), fertiliser and pesticides applications in bullock farms while in other farms preparatory cultivation, threshing, transportation also involve the use of commercial energy.

3. In bullock operated and tractor operated farms, total energy consumption is very less in medium farms than that of the other categories surveyed.

5.2.4 Output Energy from Paddy

Table 5.3 indicates the output energy in paddy production per acre of various categories of farmers. The grain-residue ratio is 2:1 in paddy cultivation in the study area. Among these farms, high amount of total output energy is produced by large farmers of bullock operated category, (7662 kWh/acre) whereas in medium farmers of tractor type, produce less amount of total energy i.e. 6991 kWh/acre.

In terms of production of total output energy, no farmer exceeds the large farmers of bullock farm. Category-wise, the large farmer has produced high amount of output energy.

Among various categories of tractor operated farms, grain production is high in large farmers (2188 kg/acre). In total energy production, the large farmers register a high level with 7554 kWh/acre, whereas medium farmers produce less amount (6991 kWh/acre) of energy. But it is still lesser than that of the large farmers of bullock operated category.

The findings of the Table 5.3 are:

1. In all categories of farms grain production is low in medium farm and the residue is high in large farm. From this, we conclude that large farmers are using more chemical fertilisers and pesticides compared with marginal farmers; over utilisation of the above inputs leads to residue growth but bio-energy use leads to high amount of grain production. Marginal farmers are using more amount of bio-energy than other categories.

2. In the total energy production, bullock operated farm leads other farm; but at the farmers' level, large farmers are leading in bullock and tractor operated farms.

3. When we compare the production (output in terms of energy units) level of all farms, medium farmer of tractor operated farm is at the lowest (6991 kWh/acre).

5.2.5 Energy Productivity, Intensity and Efficiency of Paddy Production

The productivity (kg/kWh), intensity (kWh/kg) and efficiency (output/input) of the energy are shown in table 5.3. The energy efficiency has been calculated for both grain and residue of paddy with total inputs used per acre. In tractor operated farms, the efficiency is high. Among all the farms, the small farmer registered high productivity and efficiency in paddy cultivation. It must be clarified that energy efficiency, energy productivity and energy intensity are based only on the appropriate and timely application of various inputs (e.g., fuel, electricity, fertiliser, chemicals, seeds, machinery, farm yard manure, human and animal labour).

Table 5.3 in the appendix-II shows the following inferences:

1. Among the entire farms, small and medium farmers register high energy efficiency and productivity than other categories of farms.

2. Bullock operated farm registered low energy efficiency (output to input ratio) ranging between 0.72-0.80, whereas other farms like tractor operated farm the energy efficiency is more than 0.82. The reason for the low efficiency in bullock farm can be traced to the high efficiency it leads to high degree of degradation of fertile land.

3. This table clearly shows that, productivity did not show much difference in all types of farmers, in the tractor operated farms. It suggests that the optimum mixing of all the inputs lead to better productivity.

5.3 Sugarcane (Saccharum Officinarum)

India is currently engaged in planned agricultural development. Sugarcane is cultivated in India since about 2000 AD. At present the crop covers about 3.4 million hectare (1.76% of the agricultural land of the country) both in sub-tropical and tropical regions of the country (Economic Survey, 1991-92).

Sugarcane is an intensive crop, which requires high-energy input, in terms of irrigation, fertilisers, chemicals and extensive use of machinery. Like paddy, sugarcane is also grown under irrigated conditions. Table 5.4 shows the source-wise energy requirements in sugarcane production per acre among various categories of farmers. From table 5.4, it is found that the application of insecticides and pesticides is negligible in all categories of farms. Thus, it can be concluded that, most of the farmers are practising biological pest control methods, which are environment friendly. For example, to control the internode borer, biological method control gives best result compared to chemical application; biological agent used to control the internode borer is Trichogramma Chilonis, which release egg parasite at the rate of 2.5 cc/release/ha. Six release for every fortnight starting from fourth month onwards will be necessary.

The utilisation of animal power and wood instruments is mostly nil in tractor farm, and there is a possibility of reducing bio-energy use in tractor operated farm. As is in the case of paddy, in sugarcane production also the inputs like seed, FYM, fertiliser energy contributes more to total energy consumption followed by seed input in all the operated farms under study.

In all the farms, total commercial and non-commercial energies of various categories of farmers are calculated to be 14327.89, 16125.60, 17087.43 and 15235.99 kWh/acre respectively in bullock operated farms and 16040.55, 15676.98, 16629.26 and 14777.83 kWh/acre respectively in tractor operated farms. It is important to note that the total energy consumption of large farmers is lower than that of other farmers.

The inferences drawn from table 5.4 are as follows:

1. Seed input use is high i.e., more than 41 per cent of the total input energy in sugarcane production. From this table we note that, among the seven crops studied under irrigated and rainfed conditions, sugarcane cultivation register high level of seed input use.

2. FYM also contributes more to the total energy used ranging between 16.39 and 24.43 per cent. Large farmer has practised high amount of chemical fertiliser followed by small farm, but marginal farmer use less amount of chemical fertiliser input than all other farmers.

Table 5.5 shows the operation-wise energy requirement for sugarcane production. Among various categories of farms, seed input and planting energy contribute more to the total energy input followed by manuring and fertilisation. Among all operations involved in sugarcane production, pesticides application (plant protection) as energy input is negligible in all categories of farms. As far as total energy consumption by various categories of farms is concerned the farmers consume more energy like 14327.89, 16125.60, 17087.43 and 15235.99 kWh/ acre in bullock operated farm. But in case of various categories of farms surveyed, the medium farm consumes more energy followed by small farms in BOF.

This table also throws light on the fact that most of the operations consume more or less the same amount of energy in all categories of farms. The operations are seed input and planting, irrigation, intercultural operation, fertiliser application, plant protection, harvesting and manure application, but the remaining two operations (preparatory cultivation and transport) have made considerable change in the energy utilisation in these farming systems.

The major inferences drawn from table 5.5 that are various categories of farmers of the sample study are practising different input energy for producing considerable cane yield per acre. If the farm of various categories from different agriculture systems fill the input energy gaps, there will be a possibility to attain good energy productivity, energy intensity and energy efficiency as are obtained in experimental farms of the research studies of various universities.

5.3.1 Output Energy from Sugarcane

Table 5.6 explains the acre basis output energy of sugarcane from four categories of sample farmers. Large farmers achieve more cane yield as well as leaves and tops among all farms. For energy accounting purpose, the weight of the leaves and tops are considered as 30 per cent of the cane weight, so that, when the cane yield increase per acre, there is a corresponding increase on leaves and tops. Apart from the large farmer, small farmers produce high amount of output energy, which is 42201 and 41923 kWh/acre, it is followed by medium farms 39526 and 38220 kWh/acre respectively in all the two type of farms of study respectively.

The yield of cane ranges between 24.92 tonnes to 35321 tonnes/acre and that of leaves among the four categories in all farming. In Tamilnadu, tops are usually burnt in the field. In some places villagers use them for roofing and also for parboiling paddy and other purposes.

The productivity (kg/kWh), intensity (kWh/kg) and efficiency of the sequestered energy are shown in table 5.6. The energy efficiency of both cane and leaves and tops of large farm ranges from 2.87 to 3.18 in all farms whereas the efficiency of different categories of sample farms ranges from 2.24 to 3.18. The energy productivity for cane plus leaves and tops ranges about 1.55 to 2.46 kg/kWh and the energy intensity ranges from 0.41 to 0.65 kWh/kg among different categories of sample farmers.

This table further indicates that the energy efficiency, energy productivity and energy intensity are based only on the sequestered energy (it may be both commercial and non-commercial energies). And, these sequestered energies are also responsible for maximum use of solar energy by the crops. The solar energy is available in abundance free of cost in tropical country like India, so that all farmers should exploit solar energy as much as possible though the timely and recommended amount of input use. But, solar energy is not taken into account for energy analysis in the study.

In a summary way the analysis reported by the tables 5.4, 5.5, and 5.6 in the appendix-II, the following inferences may be drawn.

1. The greater portion of the total sequestered energy (more than 41 per cent) is consumed for planting (seed input) among various categories of farms (table 5.4).

2. High energy (9.99 per cent to 23.78 per cent) is spent to mannuring sugarcane crop.

3. Considerable energy is also spent to transport sugarcane to the processing factories (6 to 8 per cent). These energy demands are due to mainly the distance of market places (40-60 kms) the mode of transport and high crop yield (e.g. for small farmers of bullock operated farm, nearly 8.61 per cent of input energy is used for transport). From table 5.5 it could be concluded that, as far as sugarcane transport is concerned, mode of transport would be tractors or lorries, which consumes less energy than bullock carts.

4. Fertilisers application (including human energy) accounts for large portion of input energy (ranges 13.34 per cent to 16.28 per cent) which is less, compared to paddy and cotton. Agro-chemicals (herbicides, insecticides and fungicides) used comprise almost minimum i.e., 0.57 per cent to 0.95 per cent of the energy demand (table 5.4).

5. Among various operations involved in the sugarcane production, planting (seed input), transportation, irrigation, organic manures and chemical energy are the operations with highest energy demands.

Energy savings can be obtained without significant yield reduction, mainly through proper use of irrigation water, fertilisers and effective use of animate energy.

5.4 Cotton (Gossypium Herbaceum)

Cotton is grown in almost all parts of the country, but it is produced in the black cotton soil areas and also to a considerable extent in all the states. In Tamilnadu, cultivation of cotton and its manufacture into textiles has been practised since pre-historic times.

5.4.1 Energy used in Irrigated and Unirrigated Cotton

The impact of irrigation on agricultural productivity and growth is enormous. The expansion and improvement of irrigation facilities occupy a prominent place in India's programmes for agricultural development (Dhawan, K.C. 1990). In this context, water (the crucial input for crop production) is a scarce commodity in India particularly in South India under dry land agricultural conditions. Thus the farmers should use water judiciously and optimally by following different techniques of irrigation. Many studies have been shown that irrigation consumed maximum amount of energy for cultivating a crop. Irrigation alone, consumed (459.11 kWh/acre) out of total (602.43 kWh/acre) required for raising sorghum crop, whereas in the case of cotton, irrigation operation used 314.98 kWh/acre out of 489.88 kWh/acre (AICRP. 1980). In the 1980s, some studies have reported that irrigation alone uses more than 60 per cent of the total energy needs for raising a crop (Batty J.C. et al, 1975). Moreover, the major share of energy consumed by irrigation comes from fossil energy, which is non-renewable (Mittal, et al., 1985).

In the study area cotton is grown in rainfed condition in larger area. Dryland cotton purely depends on rainfall, which is the only water source. Irrigated crops require more energy than rainfed crops. Under garden land conditions, electrical energy is the major source of energy and its share in the total energy is as high as 90 per cent in the farm as a whole. But its share ranges from about 75 per cent to about 95 per cent among major irrigated crops. Under rainfed conditions, bullock is the major source of animal energy followed by human energy. The share of bullock energy in the total energy is about 67 per cent and the share of human energy is about 32 per cent (Rajendran, et al., 1987).

The factors contributing to yield instability are many in dry farming. Among these are the climatic factors like rainfalls are important factors behind yield viability. A recent study suggests that agricultural output has become more sensitive to rainfall in the post-green revolution period, with the elasticity of output with respect to rainfall increasing for most crops in this period (Rao et al, 1988). This is because of the strong complementary relationship between use of modern inputs and availability of moisture either through rainfall or irrigation as also the fact that growth has taken place under diverse environments and constraints. Other empirical evidence (Nadkarni and Deshpande, 1982), however, revealed that rainfall could not wholly account for the increased variability, which depends on other factors too.

Table 5.7 explains the source-wise energy requirements in cotton production in both irrigated and unirrigated farms of various categories. From this table we conclude that energy consumption is more in irrigated farms than the unirrigated one irrespective of crops. In the two types of farming operation, bio-energy i.e., non-commercial energy consumption is high in all categories and all crops, whereas in the case of commercial energy use (bought energy), bullock farm consumes less amount of energy than the other farm under both unirrigated and irrigated conditions. It was observed that bullock operated farms consume greater share of bio-energy for different operations while tractor farms consume greater share of commercial energy than other farm i.e. bullock operated. Share of diesel and petrol oil consumption continuously increased with the increase in size of land holdings.

It is clear from table 5.7 that the per acre energy consumption of different components of energy varies considerably. Per acre total energy consumption of bullock-operated farm ranges between 8106.54 kWh to 9082.85 kWh, while for the tractor-operated farms it was 9679.70 kWh to

11523.39 kWh in irrigated farms. A very interesting point to note that, marginal farmers of bullock operated farms consume less amount of energy, which mean that they mainly depend upon animal energy for cotton cultivation.

It is observed during the examination of pattern of energy consumption of different components of the energy that per acre consumption of electricity was highest (288.60 kWh) in case of marginal farmers and lowest (222 kWh) in case of large farmers. On the contrary, per acre diesel and petrol consumption of marginal farms were lowest (78.16 kWh, and 1667.35 kWh in bullock, tractor operated farm respectively) and it was highest increase of medium farmers (133.60 kWh, and 1845.81 kWh) in two types of farms respectively. In bullock operated farm, the per acre diesel oil consumption dropped from 133.60 kWh for medium farmers to 78.16 kWh for marginal farmers. Again it may be due to the lower intensity of ploughing and irrigation on the marginal farms.

5.4.2 Operation-wise Energy Consumption

In cotton production, as we know, energy is consumed mainly for different purposes viz., preparation of soil, seed input and planting, irrigation, manure and fertiliser application, intercultural operation, plant protection, harvesting and transport (for farm purpose only). The operation-wise distribution of total energy consumption with respect to farm operation and farm size wise are discussed in table 5.8 for irrigated and dry farming cotton.

It can be seen from Table 5.8 that the total energy consumption for various operations in cotton production per acre is different, and it depends upon the operations followed and farm sizes. Manure application contributes more energy to the total energy consumption (above 35 per cent) in irrigated and dry farming cotton consumes more (more than 55 per cent). Apart from manure application, it is clear from the table that maximum energy is consumed for fertiliser application and transport and minimum energy for harvesting and seed input and planting in wet farming (table 5.8). On the other hand, intercultural operation and mannuring consumed more percentage of energy while transport and seed input consume less energy in rainfed cotton. It is noticed that, total energy consumed by bullock operated farm ranges between 8106.54 kWh to 9080.85 kWh, tractor operated farm ranges from 9679.70 kWh to 11523.39 kWh in irrigated cultivation. And it ranges from 4330.36 to 5572.94 for bullock-operated farms and 4521.89 to 5922.05 kWh/acre in tractor farms of rainfed cotton.

It is evident from table 5.7 that the total energy consumption is less in dry farms than in irrigated farms. But in the case of bullock operated category, both irrigated and dry land cotton farms consume less amount of total energy than other farms like tractor operated farms. Electricity consumption is nil in dry farms while in irrigated farms it is 2.47 to 4.85 per cent to the total energy consumption.

Tables 5.9 explain the output energy of various crops, by operation-wise and category-wise. In irrigated farm, the output energy is higher in large farm (10078 kWh/acre) and lower in marginal farm of bullock operated category (6955 kWh). While in dry land cotton, output energy is also higher in large farm (6240 kWh) and lower in marginal farm (3759 kWh) of bullock operated category. Generally, these tables indicates that large farmers are getting more output cotton than other type of farmers in both irrigated and rainfed conditions. There is no much difference of outcome energy for large farmers under dry land category are found from the table 5.9.

The residue ratio is high in marginal farm of irrigated cotton in all the three types of farms – bullock and tractor farms. In dry land cotton the residue ratio is 1:1.25 in all categories of farms.

These are the inferences of tables 5.9.

(i) In all categories of irrigated cotton, cotton lint production is high in large farms and residue production is maximum in medium sized farm of tractor operated farm, but in bullock farms, large sized farms registered high amount of both cotton and residue production. From this, we conclude that large sized farms of the bullock operated farms use high amount of bio-energy but they also add optimum amount of fertilisers, so that they achieve high amount of both cotton lint as well as residue production.

(ii) In total production of cotton, large farmer leads in all other categories of farms.

(iii) When we compare the energy production level of all farms, bullock operated farm is at the top in irrigated condition. If the farmers use bio-energy at maximum level with minimum quantity of fertilisers and pesticides they can achieve maximum cotton production.

(iv) In rainfed areas, the energy production in tractor operated farm is high; that is due to the high amount of residue yield. It leads to achieve high total energy in this farm.

The productivity (kg/kWh), intensity (kWh/kg), and efficiency of the sequestered energy are shown in table 5.9 (irrigated and rainfed cotton). The efficiency has been calculated for both cotton (lint with seed) and cotton sticks, i.e. total yield of cotton cultivation. In two different types of farms studied, bullock farm registered high efficiency and productivity in all farms than other tractor farms in irrigated cotton. But in dry farming, large farms in tractor and bullock farms have achieved considerably high efficiency rate. Marginal and small farmers achieved equal rate of productivity as in the both farming operations. This clearly indicates that the energy efficiency, energy productivity and energy intensity are based on the sequestered energy and total yield received (the sequestered energy may be both commercial and non-commercial energies).

From the data received from tables 5.9 for irrigated and rainfed cotton production, the following inferences could be drawn.

(i) High energy efficiency and high energy productivity have been registered by bullock operated farm in the case of irrigated conditions, while in the case of dry farms large farms achieve high degree of the energy efficiency in bullock and tractor operated farm. But in the energy intensity, medium farmers and marginal farmers dominant in tractor operated and bullock operated farms respectively in dry cotton.

(ii) In the case of irrigated cotton, bullock operated farm registered high amount of energy efficiency, whereas other farms like tractor operated farm have low energy efficiency. The reason for the high-energy efficiency in cotton in bullock operated farm is utilisation of maximum amount of bio-energy with optimum use of fertilisers and other agro-chemicals (it is reverse in the case of paddy). This type of production increases the land fertility, which leads to an environment friendly and sustainable agriculture. In case of dryland cotton, the efficiency of bullock operated farm is lower compared with other farms; here bought energy use exceeds the limit, it leads to high degree of degradation of fertile land.

5.5 Groundnut (Arachis Hypogaca)

In India oil seeds are grown in an area of about 20 million hectares, of which nearly 84 per cent is in rainfed condition. India is the third largest producer of oilseeds in the third world.

Table 5.10 gives the source-wise energy required for producing groundnut in irrigated and dry farming. This tabular analysis provides per acre consumption of different components of energy and the total energy consumption as regards to farm size and its comparison with farming operations. It is obvious from table 5.10 that deviations in the use of different energy input for groundnut production among various categories of farms. However, these variations may be due to the difference in the size of farm units and the availability of different components of energy.

In bullock operated farm, non-commercial energy consumption is higher. Here, Farm Yard Manure (FYM) contributes more (more than 60 per cent) followed by animal energy to the total energy consumption. Even though the non-commercial energy consumption is high bullock operated farm, the commercial energy consumption is considerably high in tractor operated farms. There is a sharp increase in the use of commercial energy in the agriculture sector (tractor operated farm) after the advent of green revolution. The consumption of fertilisers, diesel and machinery has increased manifold in the country. The total consumption of commercial energy increased from 1.8 million tonnes (calculated as coal replacement) in 1953-54 to 33.3 million tonnes in 1990-92 an average annual growth rate of 65 per cent (Palwinder Singh 1992).

In the total energy consumption, the tractor operated farm consumes less amount of energy ranging from 5117.73 kWh/acre to 5882.59 kWh/acre by large farmers and marginal farmers respectively in irrigated groundnut. Within this small amount of total energy consumption, commercial energy contribution is more than in bullock operated farms. In tractor farms the average commercial energy utilised is 23.10 per cent whereas it is about 12 per cent in bullock operated farms respectively. In the case of dryland farms the situation is same (16.64 and

5.39 per cent respectively) as in the irrigated farm but the percentage contribution is rather minimum. Through this analysis, we conclude that environmental and agricultural sustainability is effected by commercial energy in intensive tractor operated farm.

For groundnut cultivation, energy is used mainly for nine purposes compared to other crops like paddy, sugarcane and cotton. The operation-wise distribution of total energy consumption is shown in table 5.11. For manuring, preparatory cultivation, fertiliser application, plant protection and harvesting, the bullock-operated farmers are using more amount of input in all categories of farms both irrigated and unirrigated groundnut. In case of tractor operated groundnut, all the categories of farmers like marginal, small, medium and large farmers use less amount of energy compared to the bullock farm. But for some inputs application large farmers exceeds the limits set by irrigated farmers.

In all the two types of farms, manure application contributes more, which is followed by transportation of agricultural inputs to the field and outputs to the market and storage places, to the total energy consumption. Energy output of groundnut production is dealt in the table 5.12 under irrigated and rainfed farming conditions. For calculating energy content of the output, energy equivalences are used, in the form of kWh unit (6.93 kWh/kg is used for kernel and 4.83 kWh/kg has been used for groundnut shell and residues). Total output energy is higher in large farmers of both type (5215 and 5098 kWh) and is lower in marginal farm (4801 and 4578 kWh) of irrigated bullock and tractor operated land respectively. But in case of dryland groundnut, there is some deviation in the total output energy consumed among different categories of farmers as is in the irrigated farm. But the gross energy production is considerably less than that in the irrigated farm.

The following are the inferences drawn from the tables 5.12.

(i) In irrigated groundnut, the total output energy (production wise), the bullock farm leads tractor-operated farm. But in case of dry land, medium and large farmers lead in bullock operated farms, where it is 5087 kWh/acre and 5354 kWh/acre respectively.

(ii) When we compare the energy production level of all farms, bullock operated farm is at the top in both irrigated and rainfed groundnut. This leads to the conclusion that, if farmers use bio-energy at maximum level with optimum amount of purchased fertilisers and pesticides, they can achieve maximum amount of kernel as well as residues in groundnut production.

Table 5.12 further, illustrates the energy productivity, energy intensity and energy efficiency of groundnut under irrigated and dry farming conditions. Here also, the efficiency has been calculated for both kernel and groundnut shell and residues, i.e., total yield of groundnut. In different types of farms studied, the large farm registered high efficiency and productivity in irrigated groundnut, while in dryland, medium farm leads in efficiency and productivity in all operated farms. In the case of energy intensity, marginal farmers of both operation in irrigated land is high; on the other hand, large and marginal farmers are leading in unirrigated groundnut cultivation. Thus, the table clearly indicates that the efficiency, intensity and productivity are based on the sequestered energy and total yield received and also the optimum mix of non-commercial and commercial types of energy.

From the table 5.12 for irrigated and rainfed groundnut production, the following inferences may be drawn.

In the irrigated groundnut, tractor operated farm registered high amount of energy efficiency ranging between 0.78 to 1.00, whereas in dryland groundnut, medium farmer has registered high efficiency of 1.06 and 0.95 in tractor operated and bullock operated farms respectively. In dry farms, all other category farmers of bullock operated farm have lower energy efficiency than all categories of farmers of tractor operated farms.

5.6 Sorghum (Sorghum Bicolor)

The development of coarse has assumed a greater importance since it directly benefits the small and marginal farmers. These crops are mostly grown under rainfed conditions. An all-time high and a leap in production have been recorded. The total production exceeded upto 36.8 million tonnes in 1993, i.e. 10 million tonnes more than the previous year, and two million tonnes more than the highest recorded earlier in 1989-90 (Venkatramani, G., 1994). Generally, in Indian Agriculture, energy consumption is varied according to the land sizes and levels of modernisation.

Table 5.13 explains the source-wise energy requirements for sorghum production under irrigated and dryland conditions. It is clear from table 5.13 that per acre energy consumption of different components of energy varies widely.

According to Anwar Alam (1994), unirrigated sorghum in Madhya Pradesh used 2923 to 3523 MJ of energy per hectare with 33 to 36 per cent derived from men, 43-51 per cent from animals and only 6-10 per cent from commercial energy sources. Contrary to this in the case of tractor farms where commercial energy accounted for 81.3 per cent. In Madhya Pradesh, weed control is the most energy intensive operation followed by seedbed preparation, harvesting and threshing. In the study area, the commercial energy consumption is somewhat low in bullock operated farm of dry land sorghum.

Though the animal energy and wood instruments used is zero in tractor operated farm, the total energy consumption shows a marginal decrease compared to other farms like bullock operated farms. In the case of total energy consumption, marginal farmer of tractor operated farm consumes less amount, which is 5411.84 kWh/acre, whereas large farmer of bullock operated farm consumes more energy (6521.75 kWh/acre) in irrigated land. In bullock operated farm, different commercial energy consumption ranges between 27.83 kWh to 271.27 kWh/ acre in dryland sorghum. Here, among the different categories of the bullock operated farms the marginal farmer consumes lowest level of commercial energy.

Energy requirements per acre, operation-wise, with respect to farm size are discussed in table 5.14. It is obvious from these tables that in the per acre energy consumption, farmyard manure contributes more in both irrigated and dryland sorghum. Apart from FYM, fertiliser application contributes more in irrigated farms ranging between 584.27 kWh to 791.18 kWh/acre. While in case of unirrigated farm, preparatory cultivation contributes more, which ranges from 349.67 kWh to 496.27 kWh/acre in bullock farm, 180.86 kWh/acre to 258.58 kWh/acre in tractor farm. It was observed that the electricity consumption is nil in dryland agriculture and use of wood instrument in tractor operated farms of sorghum cultivation is also negligible (table 5.13).

In irrigated farm, irrigation schedule is adjusted according to weather conditions and upon the rainfall in the study area. In order to minimise the irrigation requirements of the different farms, regulation of irrigation should be followed according to the growth phases of the crop, such as growth phase (1 to 40 days), flowering phase (41 to 70 days) and maturity phases (71-95 days). Table 5.15 shows that output energy from sorghum production under irrigated and unirrigated conditions in all categories of farms. As far as the total output energy is concerned, large farmers are leading other categories like marginal, small and medium farmers in both irrigated and rainfed sorghum. Residue yield is higher in sorghum than other field crops.

In that output energy production level, large farmer of bullock operated farm is at the top in irrigated farm whereas in unirrigated farm, marginal farmers produce less energy than tractor farms. From this the conclusion made is that, bullock (livestock) is indispensable for Indian agriculture. Ramaswamy (1993) reveals that, productivity of India's agriculture sector is dismally low at about 1.2 tonnes/ha when compared to China's 3.5 tonnes/ha. China produces 250 MT of grains from 100 million hectares of land, while India is able to produce only 175 MT of grains from 150 million hectares of land. The National Commission on Agriculture has estimated that energy input has to be doubled in order to obtain increased agricultural output. Such a massive requirement of energy input cannot come from mechanisation and therefore, drought animal power has to remain the main source of rural energy for cultivating land.

Table 5.15 indicates the productivity, intensity and efficiency of sorghum production in all categories of farmers under irrigated and unirrigated conditions. Here also, the efficiency has been calculated for grain and residue i.e., total yield of sorghum. In the different types of farms studied, large farmer of TOF registered, high efficiency (3.71) in irrigated sorghum, while in unirrigated small farmer of tractor operated farm registered high (4.57). From this we conclude that, energy with other machineries practised in farm activities, give good yield. As stated earlier, marginal farmers under tractor operated farm, have cattle, which are not used for draught purposes, but used for manure purpose (dung and other animal wastes).

Inferences drawn from the tables in appendix-II under Sorghum are:

(i) Sorghum produces more output energy than other crops next to sugarcane. Residue contribution is also high, which is used for feed, fuel and other purposes in the study area.

(ii) Total energy use in Sorghum is higher in irrigated farm and is very low in dryland sorghum.

(iii) Various types of commercial energies involved in different field operations are high in sorghum cultivation under irrigated conditions. The increase is almost two times higher than unirrigated farm.

5.7 Bajra (Rennisetum Glacum)

Bajra is one of the important cereals in India especially in Tamil Nadu. Table 5.16 explains the source-wise energy requirements in Bajra production both irrigated and rainfed farms various categories. In the total energy consumption of the irrigated bajra cultivation, the lowest is recorded by small farmer of tractor operated farm (5133.15 kWh/acre), the highest is achieved (6401.49 kWh/acre) by large farmer of bullock operated farm. While in rainfed bajra the total energy consumption is very less, the highest range is 5814.05 kWh/ acre by large farm of bullock operated farm and lowest range is 4668.47 kWh/acre by marginal farmer of tractor operated farm. From this we conclude that, energy consumption is more in irrigated farms than in the unirrigated one, irrespective of farm sizes.

Tractor operated farm consumes less amount of total energy than other farm; but contribution of commercial energy here is more to the total energy. In the same way bullock operated farm registered high in total energy consumption, but contribution of commercial energy is less, which does not create any problem in agro-ecosystem.

One can observe from Table 5.17 that total energy consumption for various operations in Bajra production per acre is varied, which purely depends upon the operations followed and farm sizes. Among various operations involved in production of bajra, manure contribute more, followed by the chemical fertiliser and pesticide application, in both the farms (irrigated and rainfed). It is important to note that, energy for seed input and planting shows huge difference (more than four times of energy) between the irrigated and the unirrigated bajra. This is due to transplanting of seedlings, which involves high amount of human labour in irrigated farm. Whereas in dryfarm direct boardcasting of seeds is followed by sample farmers in the study area, here one labour can do the job (per hectare) in one hour, so that, it requires less amount of energy. Moreover, dryland crops require more energy for interculture operations due to more of weed growth, because the land is cultivated once in a year only (monocrop).

Output energy of bajra production has been dealt in the tables 5.18 of different farming like irrigated and irrigated unirrigated systems respectively. It is obvious from the table that, the grain residue ratio of irrigated bajra is 1:2 and that in unirrigated farm is 1:1.25. For conversions from physical units to energy units, energy equivalence table are used as for instance 4.07 kWh/kg is for bajra grain and 5.0 kWh/kg is for its residue. In the total output energy, the energy production is high in large farm of TOF type (16090 kWh/acre) whereas it is less i.e., 12936 kWh/acre in marginal farm of bullock operated farm for irrigated bajra. In the case of bajra grown in rainfed conditions, the output production is high in large farm (7948 kWh/acre) of bullock operated system and it is less in medium farm of tractor operated farm (5225 kWh/acre). From this analysis, one conclusion is that, for good returns (yield) from bajra production, irrigated farm required optimum mix of both commercial and non-commercial energy while unirrigated farm requires more amount of bio-energy, so that high amount of yield is registered in bullock operated farm.

Table 5.18 depicts the energy productivity, energy intensity and energy efficiency of various categories of farms with operation-wise differences and also dealt with the mode of irrigation of bajra production. Output energy production is increasing correspondingly with the increase of input.

Highest energy efficiency is registered in small farmer category (2.77) and low efficiency (2.38) in marginal farm of bullock operated irrigated farm, whereas in dryland bajra large farm of bullock type and tractor type registered high amount of energy efficiency. The small farmers of bullock operated farm and medium farmer of tractor operated farm has high productivity (0.59) in irrigated sorghum.

The following inferences could be drawn from the data available in tables related to both irrigated and unirrigated bajra.

(i) High amount of total energy is registered in bullock operated farm, eventhough the contribution of commercial energy is less. The contribution of commercial energy in the bullock operated farm is the least among the unirrigated types of farms.

(ii) Energy differences in seed input and planting in both irrigated and unirrigated bajra are due to human labour involved for transplanting in irrigated farm and direct seed broadcasting in unirrigated farm.

(iii) Irrigated bajra requires optimum mix of commercial and non-commercial energy, whereas the dryland required only bio-energy with minimum of commercial energy.

5.8 Pulses

There are number of types of pulses cultivated in the study area. They are redgram (canjanus cacan), blackgram (vigna mungo), Greengram (vigna radiata), cowpea (Bigna unguiculta), and horsegram (macrotyloma uniflorum). Energy analysis for pulses is the average value of the all grams.

Table 5.19 explains the source-wise energy requirement of pulse production of irrigated and dryfarm. Comparing with various categories of sample farmers, large farmers consume more than 5229.78 kWh/acre of energy input per acre of total energy consumption. It is obvious from this table that the fertiliser consumption is nil in the sampled farms except large farmers (they too use a little amount). There is considerable difference in seed input and planting in both farm (irrigated and unirrigated) as shown in table 5.20. The energy consumption for these operations is less in irrigated farm and more in unirrigated farm. Generally seeds are broadcasting in both the farms, but an excess amount of seed is used for dry farming (due to uncertainty in it). In the total energy consumption, small farmer of bullock operated farm consumes less (4303.76 kWh/acre) while it is more in large farmer (5229.78) of bullock operated farm. In dryfarm, it is high in large farm (5494.48 kWh/acre) of bullock operated type, but less in marginal farmer of tractor operated category (3968.90 kWh/acre).

The table 5.21 explains the output energy of pulses. In the total yield of pulse, grain-residue ratio is very low. The energy equivalence used for energy conversion of yield is 4.08 kWh/kg for gram and 3.47 kWh/kg for pulse straw.

As far as the total output energy in irrigated farm is concerned, maximum output is produced by the marginal farmers of tractor farm (3805 kWh/acre) than the others. The output energy is very low in dry farming pulses production. It is explicitly explained by the table 5.21 that there is no corresponding increase in the yield with increase of each unit of input, in all categories of farms. High range of efficiency (0.76) is registered in marginal farm of TOF (irrigated), and low efficiency is seen in large farm of tractor operated farm (0.45) in unirrigated land.

In the case of energy efficiency, various categories of farms show better efficiency than the large farmers of rainfed cultivation, because they get considerable yield with minimum of input. The following are the inferences drawn from the tables discussed above under pulses production.

(i) Fertiliser energy consumption is nil in marginal, small and medium farms, but large farmers are using some amount of fertiliser, which is very minimum compared with fertiliser utilisation by the other farmers.

(ii) Farmers use more amounts of seeds in dry farming than the irrigated farm. This is mainly due to unfavourable conditions for germination of seed under dryfarming condition.

(iii) Total output energy is high in marginal farm (TOF) for irrigated case and medium farmer is high in the case of dryfarm pulses production.

(iv) There is no corresponding increase of yield with the unit increase of input in all types of farmers.

5.9 Summary

Energy is widely viewed as a "basic need" in the same way as food, clothing and shelter. In fact, energy for agriculture sector may be considered as important as origin of the food production. Moreover, poor farmers spend a relatively higher proportion of their income on fuels and fertilisers than the rich famers. Consequently, increase in price affect those small and marginal farmers proportionately more. In the case of pattern-wise farming, irrigated farmers are using more energy than unirrigated one in the proportion of the total energy used for different crop cultivation. Hence, it is necessary to consider policy measures, which can meet the energy needs of the agriculturists.

6

The Influence of Energy Input

6.1 Introduction

The influence of various agricultural inputs on yield are determined by different factors such as the size of the landholding, types of operation (bullock and tractor farm) mode of cultivation (irrigated, unirrigated), land and soil types of different villages and also various types of crops. This chapter is an attempt to study the energy inputs, which significantly influence the yield or output. In this chapter, the results of the energy consumption models, with respect to the whole sample, crop-wise, operation-wise and farm size-wise, have estimated are presented and discussed. The tables are shown in the appendix-III.

6.2 Village-wise Energy Use Pattern in Agriculture

The table 6.1 provides the village-wise and source-wise energy use pattern for agricultural operations for the year 1997-98. The total number of village panchayats covered in the survey is 12. Out of 240 total number of plots surveyed, irrigated plots are 60 and 180 plots are unirrigated. As a whole, in T. Kalluppatti block, 240 farms consume 54.72 lakh kWh of energy per annum, to cultivate crops in 479.95 acres of land.

Out of twelve sample villages in T.Kalluppatti block, Pulkattai village panchayat ranks first in gross sown area and Kunnathur village panchayat first in agricultural energy consumption. Kunnathur consumes 5.27 lakh kWh of energy per annum for cultivating crops in its 46.4 acres of land. The second position goes to Genjampatti panchayat. Its agricultural energy consumption is 5.19 lakh kWh in its cropped land of 36.3 acres. If one divides the total energy consumption level into different sources of agricultural energy, this table explains that, energy consumption of every village is influenced by different type of energy used during the year 1997-98. This may be due to the following reasons; (1) larger gross area sown, (2) larger availability of levelled land and absence of undulating topography; (3) modernised agricultural practices etc.

Comparing Pulkattai with Kunnathur, the area under cultivation is high in Pulkattai but it consumes low level of agricultural energy, which occupies 10th rank among these 12 villages. From the sample area, Pappaiyapuram village consumes very minimum amount of machinery energy (39138.4 kWh/annum) for its crop cultivation of 33.95 acres of land, which is mainly effected its least amount of total energy utilisation (361844.6 kWh/annum) during the year of cultivation. It directly indicates the fact that dry farming and rain-fed crops consume less energy compared to wet farms and its crops.

Of all type of total energy consumed by the villages, FYM occupies first place, followed by machinery utilisation, seed input etc. Least herbicides use (750.33 kWh/annum) indicates that the farmers are not giving much importance in the study area. It is observed from the table (6.1) that, the Vannivelampatti and Velampur panchayats must have larger amount of draught animals due to its use (39914.2 and 38488.6 kWh/acre respectively) among all other villages selected. Chittulotti village is using very minimum amount of animal energy. It is clear from the table that, the farmers depend more on machinery than animal energy. It is also a fact that this village has meagre animal population.

Though seed input occupies third largest amount of energy utilisation, Genjampatti and Velampur villages use more energy followed by the villages Vannivelampatti, Kunnathur and Chittulotti. It is understand from this, soils of these villages have less germination power, due to its natural conditions. Since Velampur panchayat has second larger number of draught animals, it is obvious that usage of wood instrument also higher (36437.2 kWh/annum) during the year 1997-98.

Regarding diesel and petrol utilisation, Melapatti and Silarpatti panchayats are using more energy (42443 and 34829.4 kWh/annum) whereas Melapatti panchayat is using minimum amount of electricity energy (21157 kWh/annum). It shows that the farmers of Melapatti panchayat are concentrating more on irrigated agriculture using oil motors but Silarpatti farmers use more electric engines for their irrigation purposes.

Pesticide and insecticide applications become inevitable in Indian agriculture today. There is no exemption for the study area i.e. T.Kalluppatti block of Madurai district (TN). The farmers of Sandhaiyur panchayat using more pesticide / insecticides (37122.2 kWh/annum), though they ranked at 5th in total land utilisation. This shows that the insects attacked both dry and irrigated crops during the year of cultivation (1997-98) in Sandhaiyur.

Generally, among all the villages, Vannivelampatti and Kunnathur have larger number of electric pump sets. It is clearly reflected in the table (6.1) that these villages use more electric energy (64447.2 and 56521.6 kWh/annum respectively) followed by Kuvalapuram and Silarpatti.

Considering the machinery use in the study region, Velampur and Sandhaiyur panchayats are at the top by using more than ninety thousand kWh of energy per annum during the year of observation. We may conclude from the table that, the farmers of both Velampur and Sandhayur are having more number of tractors and machinery implements for supporting their agricultural operations.

The table 6.2 explains the energy use in agriculture as per acre basis of cropped area. Among the various energy sources, FYM energy accounted for above 25 per cent of the total energy consumed in nine villages. On per acre basis, the village (Genjampatti) was the maximum energy of 14308.31 kWh, of which FYM energy contributes 31.14 per cent, machinery 17.20 per cent, seed 14.01 per cent and fertiliser 5.97 per cent etc. This same pattern is observed in many of the sample villages.

Very interesting points emerge out when one analyses the energy use pattern per unit (acre) of cropped land (table 6.2). Out of all the sample villages in T.Kalluppatti block, Genjampatti (fourth village) ranks first in total agricultural energy consumption per acre of cropped area. The second rank goes to Velampur panchayat. It consumes 14295.37 kWh of energy per acre per annum.

The farmers of Pulkattai village consume least energy of 7270 kWh per acre per annum. This low level of agricultural energy consumption per acre once again indirectly indicates the fact that dry farms and the crops cultivated there consume less amount of energy.

In T. Kalluppatti block, on an average, the agricultural energy consumption per acre of cropped land per annum is 11401.57 kWh. As many as seven villages (except Genjampatti Velampur, Nallamaram, Silarpatti and Kuvalapuram) consume energies below the average energy consumption of the block (table 6.2).

Further, it also explains that human energy usage is maximum in the Silarpatti village (9.71 per cent) and seed energy usage is maximum in the Velampur (14.16 per cent) and Genjampatti (14.01 per cent) to the total energy consumed. It can be observed from the table that the consumption of FYM ranges from 17.71 per cent (seventh village) to 37.53 per cent by the ninth village. The total energy used is minimum in the first village where the pesticide application is also least i.e. 1.94 per cent among all the villages. By and large the major energy inputs are renewable sources like farmyard manure. The commercial energy constitutes 43.96 per cent to the total energy consumed in the study area.

The examination of the pattern of energy consumption in relation to operation wise and farm size wise is important to know the variations in the consumption of different components of energy (animate energy, bio-energy, electrical energy, chemical and mechanical energy) of the selected farm units. In fact, both operations and farm sizes may be important determinants of energy consumption. This section is concerned primarily with the examination of differences in the pattern of energy consumption of different farm operations and farm sizes of selected crops. Analysis of data in this section is based on per acre consumption of different components of energy, i.e., animate energy, bio-energy, electricity, chemical and mechanical energy.

6.3 Energy Consumption in Irrigation Farm

Table 6.3 shows a brief account of total energy consumption in kWh by selected irrigated crops in both bullock operated and tractor operated farms on per acre basis. Energy values used in this table are average energy of all categories of farmers (farm sizes). In the bullock operated farms, the sugarcane cultivation required maximum energy (51349.92 kWh) compared to other crops like paddy (31059.98 kWh), cotton (27172.82 kWh), groundnut (19203.95 kWh), sorghum (19127.14 kWh), bajra (17691.56 kWh) and pulses (15656.77 kWh).

The table 6.3 also shows that in the total energy consumption, the high percentage of energy sources accounted for different crops are such as farmyard manure 71.13 per cent for pulses; fertiliser energy 18.90 per cent for cotton; electricity 12.15 per cent for paddy; seed input 47.99 per cent of sugarcane; animal energy 10.76 per cent paddy; human energy 9.03 per cent for cotton and wood instrument 6.81 per cent for groundnut. It is very clear from the table that FYM and seed input constitutes a major share of input energy use in the production of output. Sugarcane registers high amount of seed input energy (24640.69 kWh) utilisation. It is the only crop utilising maximum solar energy (Menon et al., 1984).

The second part of the table 6.3 reveals about the energy consumption in few crops for Tractor Operated Farms. In this type of farms mechanical energy is mostly used for land preparation, threshing and transportation purposes. But there are variations in the other components of various sources of energy. The use of animal power and wood instrument are completely nil (except for paddy crop 0.24 per cent, which is required for levelling the land). While comparing the different crops by their consumption level of different sources of energy, it is found that pulses consume 72.68 per cent of FYM. 50.47 per cent of seed input consumed by sugarcane, 18 per cent of fertiliser and 13.58 per cent of electricity for paddy, 7.14 per cent of human energy and 17.44 per cent of diesel and petrol for cotton.

Among all the crops, the seed input requirements are more for sugarcane followed by groundnut in both the operations. But paddy consumes more amount of FYM (12884.95 and

12896.55 kWh/acre) for bullock and tractor operated farms. The crop cotton requires 15.79 per cent of mechanical energy (5212.2 kWh/acre) in TOF whereas, it is only 1.83 per cent BOF. Thus, the use of high or low level of different sources of energy depends upon the nature of the energy involvement in the various practices of the different crops. Electrical energy consumption is high in paddy and sugarcane cultivation, compared with other crops, because these two crops require more amount of water. For the remaining crops, the electricity consumption differs according to their water requirement and the seasons in which they are cultivated.

As far as use of fertiliser is concerned, sugarcane registers high consumption of 7121.39 and 7116.14 kWh in both operations, but on the basis of percentage of total energy, paddy accounts for 18.00 per cent in TOF, whereas cotton accounts for 18.90 per cent in BOF. The variations in percentage of fertiliser energy utilisation among sugarcane, paddy and cotton are due to differences in crop period, and with respect to other energy inputs, fertiliser extraction is very high i.e. 7121.39 and 7116.14 kWh in TOF and BOF respectively in sugarcane and very low in pulses (56.10 kWh). Chemical and mechanical energies are utilised to a maximum in tractor operated cotton compared to other crops. It can be observed from the table that FYM and fertiliser energy is the energy intensive sources in the cultivation of major crops in the study area.

6.4 Energy Consumption in Dryfarm

Major crops cultivated in unirrigated farms are cotton, groundnut, sorghum, bajra and pulses. The total energy consumption varies between the irrigated and unirrigated crops under different farm operations. First part of the table 6.4 throws light on the total energy consumption in kWh by selected crops in bullock operated farm. The most important energy intensive inputs for drycrops are FYM and animal energy. Manure energy utilisation capacity is high in sorghum (68.05

per cent) and low in cotton (59.06 per cent). Animal energy contributes 11.54 per cent in groundnut but a very low 8.29 per cent in pulses to the total energy consumption.

In the study area, dry crops are grown without irrigation and they purely depend on rainfall. So, electrical energy consumption is nil in all crops. Among the total energy consumption for all crops, energy utilisation for sorghum is high with 17912.12 kWh/acre whereas pulses consume less amount of energy (14150.65 kWh/acre). In the bullock operated farms, contribution of animal energy ranges from 8.29 per cent for pulses to 11.54 per cent for groundnut to the total energy. Therefore, it is clearly explained that groundnut is an energy intensive crop among other crops.

The table 6.4 gives a picture of the total energy consumption by the selected dry crops in Tractor Operated Farm. In contrast with the bullock-operated farms, animal power and wood instrument use are completely nil in the tractor-operated farms. Among the various dry crops here, bajra requires the total energy of 16566.52 kWh/acre compared to pulses (13620.82 kWh/acre). Here also, utilisation of FYM energy is high in groundnut (11060.35 kWh/acre) while it is low in cotton (9450 kWh/acre). In this farm, total commercial energy utilisation is high compared to bullock operated farm. Major inputs contributing more to the total energy consumption for all crops are FYM and fertiliser energy. It is also important to note that, commercial energy use is high in tractor operated farms. Hence, it warrants finding out a possible way to reduce commercial energy with marginal increase of bio-energy such as farmyard manure, green leaf manure, and bio-fertilisers and pesticides.

In the case of groundnut the various energy sources are distributed as bio-energy 70.86 per cent, seed input 8.20 per cent and diesel/petrol 7.28 per cent. The least source of energy consumed by pulses is fertiliser, which has accounted for 0.22 per cent to the total energy consumption.

While considering the total energy, the minimum energy requirements are 13620.82 and 14150.65 kWh/acre for pulses in both the farm operations. It is observed that human energy is required more on cotton cultivation than the other dry crops. It is clear from the table that during harvesting time, it requires more labour power irrespective of farm operations.

6.5 Influences of Energy Inputs on the Output of Selected Crops

Here, the study discusses the results of the output models, with respect to the whole sample crop-wise, farm category-wise, and farm operation-wise for irrigated and dry crops. This is achieved by estimating regression equations with yield as the dependant variable and type of energy as independent variables, for each selected crop.

6.5.1 Results of Crop-Wise Regression Models in Irrigated and Rainfed Farms

Table 6.5 and 6.6 analyse the regression co-efficient of crop-wise energy consumption of both irrigated and dry farms. In the regression equation for the sample farms, the value of F is statistically significant (p=0.001) and the model is valid for inferences.

The dashes present in the table indicate that the variables concerned have not been included in the regression equation to avoid multi-collinearity, resulting from near proportionate level of input energy to 'area'. They are, in fact, important inputs whose influences are exerted through the variable 'area' for all crops.

Multi-collinearity

(a) High collinearity is often encountered in agricultural economics as well as energy analysis. For example, in farm or crop production function studies, the exploratory variable viz., area, human energy, bullock energy, mechanical energy, chemical energy and bio-energy show the same broad pattern of behaviour across various farms. In consumption or demand studies, this situation can be encountered when observations on one of the explanatory variables is an exact multiple of observations on the other variable (Madnani, 1988).

(b) The co-efficient of multiple determination (R^2) may be high but majority of all of the regression coefficients may be insignificant. This is a situation, which will lead one to interpret that the explanatory variables taken together cause a very high variation in the dependent variable while no individual variable exerts any significant effect. Thus presence of multicollinearity may lead to wrong interpretation of the effect of some explanatory variables.

Area

Area is found to have positive influence on the yield in all the crops in both irrigated and rainfed farms. It is statistically significant at 0.1 percent in case of all type of crops. It is as expected result, because the larger the area, the larger will be the energy requirements.

Animate Energy

Animate energy is found to be significant in the crops, like paddy, sugarcane, sorghum and bajra but not in the other three crops such as cotton, groundnut and pulses. This variable has a negative influence on sugarcane, sorghum and bajra, because farmers are using extra animate energy, implying that an increase in the utilisation of this energy will not cause an increase in the yield but on the contrary there will be a possibility of decreasing the efficiency of animate energy. But this has a positive influence, with paddy only; with an increase of one unit of this energy, there will be an increase of 0.1 kg. of output per acre. In rainfed farm, animate energy is significant only for cotton and pulses. But this variable has positive influence in pulses, though marginal and negative influence on cotton. In case of animate energy the size of the co-efficient is very small except in the case of sugarcane. Animate energy has inverse relationship to the area cropped, in case of sugarcane, sorghum and bajra for irrigated farms; cotton and bajra for unirrigated crops. Since bajra is grown in small farms, it is likely that animate energy will be less utilised. Moreover, this regression co-efficient is statistically not significant.

Bio-Energy

Bio-energy had positive influence on three crops viz. cotton, sorghum and bajra. But in the case of paddy, groundnut, and pulses, there was a positive influence and the influence was nearly proportionate to the area. Among the crops positively influenced by bio-energy, bajra and cotton had significant levels with p=0.001 per cent level. In the rainfed farms, bio-energy had a positive and significant influence on cotton only. In case of sorghum and bajra, the level of significance is proportionate to the area. The impact of bio-energy on area cropped is not significant for sugarcane of irrigated farms, groundnut and pulses for unirrigated farms. One can conclude that bio-energy is less significant in the cropping pattern of both irrigated and dryland farming in the study area.

Mechanical Energy

Among the selected crops, only two crops, paddy and cotton are positively influenced by mechanical energy. In the case of sorghum and pulses, the influence is nearly proportionate with that of animate energy. Paddy has a significant p=0.001 level and cotton has p=0.01 level. Mechanical energy includes electricity and pump, which are highly essential for paddy, because it requires more water compared to other crops. The application of mechanical energy is near proportionate with area in dry farms. The mechanical energy has some significant impact on area cropped in irrigated farms. Its impact is statistically significant for paddy and cotton crops. In case or rainfed farms, mechanical energy has some impact on sorghum crops but it is not statistically significant.

Chemical Energy

Groundnut was positively influenced by chemical energy whereas for other crops such as paddy, sugarcane, cotton, sorghum and bajra, it is nearly proportionate to the area. The crop-wise regression coefficients of the consumption models, the chemical energy increased proportionate to the land area and other factors like weather conditions of the particular season, incidence of pest and so on.

In drylands, farmers were using insecticide and pesticide dust at the matured stage of sorghum and bajra considerably, and so chemical energy had positive influence on these two crops. Interestingly, the importance of chemical energy on area cropped has shown significant in unirrigated or rainfed farms of sorghum and bajra. The impact of chemical energy for groundnut is statistically significant for irrigated crops. It is understood that unirrigated farms use more pesticides and fertilisers, the source of chemical energy, to boost the agricultural yield, which generally depends on rainfall.

The major inferences drawn from this table are that bio-energy has dominant role compared to other energy while chemical fertilisers have some importance. Eventhough chemical energy has importance, there is a possibility to reduce the utilisation of the same without reducing the productivity. This will also help to develop sustainable farming practices, which will not affect ecology and agro-environment. The use of bio-energy has a potential influence for output increase. As indicated by the models it would be taken into consideration for future planning. In dry farming most of the farmers is not much concentrating on input application except the animate energy utilisation.

6.5.2 Results of Operation-wise Regression Models in Wet and Dry Farming

All variables involved in the two (bullock and tractor operated) farms are valid for interpretations since the values of R^2 are high in both irrigated and dry farms (tables 6.7 and 6.8). The independent variables included in the models, in irrigated farm, explains that more than 90 per cent significance in tractor operated farms and 78 per cent in bullock operated farms. This is due to the fact that most of the farmers in bullock operated farms are marginal, uneducated and poor people, thus the level of input application is moderate compared to other farms. In some villages marginal farmers belonging to bullock operated farms, are using prescribed level of inputs but they do not adhere to the time schedule of inputs application one could see the similar trend in the dry farms also.

Area: In the operation-wise energy consumption models, for both irrigated and rainfed farm, the area has negative influence on the dependent variable (yield). It is observed from these tables that when the area increased, there is no increase of yield in bullock and tractor operated farms. The co-efficient is large also.

Animated Energy: Animated energy in both irrigated and unirrigated farms has a negative influence on yield except in the unirrigated bullock operated farms (p=0.01).

Bio-Energy: In the operation-wise energy consumption regression coefficient models, the bio-energy has a positive influence with p=0.001 level of significance in all irrigated farms. Bio-energy influences considerably the level of yield in the two categories of irrigated farms only, but not in the unirrigated farms.

Mechanical Energy: This variable has significant positive influence in both farms of unirrigated condition only whereas in irrigated farms it is not significant in all the categories of farming. The significant level in bullock and tractor operated farms is p=0.001. There is a possibility to increase the output through the mechanical energy use in dry farming.

Chemical Energy: In irrigated farm, chemical energy has a significant positive influence in bullock operated farms and its level of significance is p=0.01. In the case of dry farming this variable has a significant positive influence, only in bullock operated farms under irrigated conditions. It has significant negative impact on Tractor Operated Farms. Tractor Operated Farms use less pesticides and fertilisers in irrigated land.

6.5.3 Results of Category-wise Regression Models in Wet and Dry Farms

In this model, the R^2 values are statistically significant among all categories of the irrigated farms, but in case of unirrigated farms, the R^2 values are statistically significant for marginal, small and large farmers, and this value is less in the medium farmers. So in the irrigated farms, all the models are valid for interpretations whereas in the unirrigated farms, except medium farmers, other models are statistically significant and valid for interpretation.

Area: In the category-wise energy consumption models, the area is negatively influencing the yield among irrigated medium and large farms only, but in unirrigated farms marginal and small farms are positively increasing on the dependent variable. When we increase one unit of land area (independent variable) there will be positive influence on yield except medium and large farms in the irrigated land.

Animate Energy: In irrigated farms, animate energy is statistically significant in three farms (marginal, small and large). Even though these farms are statistically significant, it has a negative influence on yield indicating over use of animate energy.

But in unirrigated farms, animated energy has a positive effect among marginal and small farms.

Bio-Energy: Among all independent variables involved in this model, bio-energy had a positive and considerable influence in all categories of irrigated farms. But it has a negative influence on marginal unirrigated farms only. In all the farms of irrigated crops, bio-energy have a significance level of p=0.001. Bio-energy has not much contributed to dry crops, because farmers have not properly followed the input applications and other sources of energy.

Mechanical Energy: Mechanical energy has a positive influence on only the large farms in dry lands whereas it has a negative influence on marginal farms of irrigated land, there will be negative effects on yield. This is due to maximum or excess use of mechanical energy in the irrigated lands.

Chemical Energy: This variable also has a positive influence on yield in marginal and small-irrigated farms but it shows negative influence in large farms of irrigated land. In dry land the marginal farms show statistical significance and negative influence on yield.

The importance of independent variables such as animate energy, bio-energy, mechanical energy, chemical energy are significant for marginal farm in the irrigated land. In the case of small farmers also these energy variables are statistically significance with the exception of mechanical energy. Overall one can say that the yield pattern in irrigated land is explained by the included independent variables. However, in rainfed lands, these energy variables do not seem to be statistically significant in all categories of farmers except for marginal farmers.

6.4 Summary

The results of the study suggest that concerted efforts are needed in agricultural sector to improve energy productivity.

Systematic energy analysis and implementation of energy conservation measures are imperative for profitability as well as sustainability in agriculture. A harmonious blend of Government policies and farmers initiatives is required for achieving the goal of energy efficient farming in the country. Bio-energy and chemical energy were the most required forms of energy for the different crops taken up for investigation. Energy savings can be achieved without yield reduction, mainly through proper use of irrigation water, chemical fertilisers, machinery and effective use of animate energy. Generally bio-energy is more promising in terms of its contribution to positive output change and extension services to help farmers to utilise bio-energy rather than commercial energy (chemical fertilisers) will go a long way in achieving eco-friendly sustainable development in agriculture.

7

Summary and Conclusion

7.1 Introduction

As with all systems, the energy system is, on the one hand, an interacting part of the economy, and on the other, organised in two parts namely energy supply and demand, which interact with each other. Energy supplies are, therefore essential both intermediate production and for final consumption in agricultural activities. But how much energy is required per unit of service performed by energy to satisfy crop production needs depends upon the technology adopted by the end user of energy in dryland. As such this chapter summarises the major findings and conclusions arrived at in the course of the foregoing discussions. Suggestions are also made for further policy formulation and research.

7.2 Preliminaries

A close observation of details furnished on energy use in agriculture production would reveal that, on the whole, in the crop cultivation under unirrigated conditions, ranking of the energy intensive unit operations will be manuring, fertiliser application, preparatory cultivation, harvesting, weeding or inter-culture operation and threshing. However, under intensive irrigated farming, the energy input ranking goes on manuring, fertiliser input, irrigation, seedbed preparation and sowing/ transplanting, plant protection and weeding or inter-culture operation. In crops like sugarcane, seed input is a major energy intensive operation. Herbicides accounts for less than one per cent of the total energy except in the case of paddy, whereas manure (FYM) constitutes more than 40 per cent in all crops of all categories.

Energy inputs in cultivation differ among various categories of farmers. The most significant among them is the level of use of modern inputs like chemical fertilisers, pesticides and different mechanisation levels. Energy use, therefore, has been determined for different degrees of mechanisation used for in each crop in the respective areas. Traditionally, and to a large extent, even now, the sources of power in agriculture were men and draft animals. Use of tractor for tillage, transport and threshing is very common now. Much more mechanised in lift-irrigation where pump sets and tube wells have virtually replaced bullock driven lifts.

7.3 Findings

The other related findings are as follows:

1. As a whole, in T. Kalluppatti block, 240 farmers spread over 12 villages panchayats consume 54.72 lakh kWh of energy per annum to cultivate crops in 479.75 acres of land. Out of the total number of plots surveyed (240), the irrigated farms were 60 and unirrigated 180.

2. Out of 12 village panchayats studied in the study block, Pulkattai village panchayat ranks first in gross sown area and Kunnathur village panchayat first in agricultural energy consumption (5.27 lakh kWh of energy per annum for cultivating crops in 46.4 acres of land). If one divides the total energy consumption level into different sources

of agricultural energy consumption of every village is influenced by different type of energy used during the year. The reasons for this can be traced to the facts such as, the larger gross sown area, larger availability of level land and absence of undulating topography and modernised agricultural practices.

3. Pappaiyapuram village consumes very less amount of energy, which is due to the minimum mechanical energy use in 33.95 acres of land. It directly indicates the fact that dry farming and rainfed crops consume less energy when compared to wet farms and its crops.

4. The energy consumption was found to increase if the area under cultivation was more only in irrigated conditions.

5. Among the various energy sources, FYM accounted for more than 27.47 per cent of the total energy input followed by machinery and seed input.

6. Per acre basis, the Genjampatti village panchayat used the maximum energy of 14308.31 kWh contributed by FYM energy (31.14 per cent), machinery (17.20 per cent), seed (14.01 per cent) and fertiliser (5.97 per cent). This is due to more application of farm yard manure.

7. In the study area, on an average, the agricultural energy consumption per acre of cropped land per annum is 11401.57 kWh. As many as seven villages (except Genjampatti, Velampur, Nallamaram, Silarpatti and Kuvalapuram) have values below the average figure, due to lesser energy consumption of the rainfed farms.

8. On the whole the dominating energy inputs are renewable sources like farmyard manure. The commercial energy only constitutes 43.96 per cent to the total energy

consumed in the study area. The level of utilisation of manure energy by field crops is purely based on the different proportional use of various organic and chemical inputs.

9. If we exclude the FYM energy in calculation of total energy, the percentage contribution of mechanical energy will be more.

7.3.1 Findings from the Crop-wise Energy Analysis

(1) In the study area, paddy and sugarcane have been grown under irrigated conditions only. Remaining crops such as cotton, groundnut, sorghum, bajra and pulses have been grown in both irrigated as well as rainfed conditions.

(2) Among various farms studied, non-commercial energy use is higher in bullock operated farm and lower in tractor operated farm.

(3) In the irrigated farms, electricity consumption contributed marked percentage to the total energy. In the study area, most of the farms are irrigated by wells and require electricity for lifting water. Among various irrigated crops, paddy and sugarcane require more water and correspondingly electricity consumption is higher for these two crops than other crops.

(4) Animate energy use is high in bullock operated farms, but is very small in tractor farms. The use of wood based instruments is negligible in tractor farms.

(5) Diesel/petrol utilisation is more in tractor operated farm than bullock operated farm for both irrigated and dry crops.

(6) In all the farms, maximum amount of energy is contributed by farm yard manure, which is more than 40 per cent of the total energy.

(7) Among production is high in marginal farm and that of residue is high in large farm. From this, we find that large farms are using more chemical fertilisers and pesticides compared with marginal farmers, over utilisation of which leads to residue growth but bio-energy use leads to high amount of grain production (marginal farmers are using more amount of bio-energy than other categories).

(8) In most of the sugarcane fields, biological control of pests give the best results compared to chemical application, the biological agent used to control the internode borer being Trichogramma chilonis.

(9) Among various crops studied, seed input energy use is high in sugarcane, which is more than 41 per cent of the total input energy followed by groundnut. As well as the FYM used is concerned, sugarcane comes at the top.

(10) Sugarcane consumes more energy for transportation, which is mainly due to the distance (40-50 km) of market place.

(11) For field and other operations the share of diesel and petrol oil consumption continuously increased with the increase in size of land holdings.

(12) Irrigated crops utilise more amount of input energy than rainfed crops.

(13) There are two types of leguminous crops cultivated frequently in the study area. They are oil seeds (groundnut) and pulses (black gram, green gram, red gram and cowpea), that enrich the soil fertility through fixing atmospheric nitrogen in the soil. Pulses consume less amount of energy than other crops.

914) Normally, commercial energy involved in different operations is higher in all irrigated farms than the unirrigated farms.

(15) Comparing to other crops, sorghum and pulses are grown mainly in rainfed farms.

(16) There was a difference in energy use for seed input in sorghum and bajra under irrigated and rainfed farming. This is due to different level of human labour required for transplanting in irrigated farm and direct seed broadcasting in dry farm.

(17) For pulses production, commercial energy i.e., fertiliser consumption is nil in marginal, small and medium farms, but large farmers are using some amount of fertilisers which is very minimum.

7.3.2 Findings from Regression Model (Influencing inputs)

In this model study, input data has been tested on output of selected crops for their influence level and their level of significance on yield both irrigated and dry farming.

(1) The major inferences drawn from this regression model are: bio-energy has a dominant role on yield compared with other energy while chemical fertiliser has also some relevance on the yield. Even though chemical energy has some importance, there is a possibility to reduce its utilisation without reducing the productivity. This will also help to develop sustainable farming practices. Another important thing is that in dry farming most of the farmers were not giving much concentration on bio-energy input application except animate energy utilisation.

(2) In the crop-wise regression analysis, animate energy was found to be significant (negative influence) in the crops of

paddy, sugarcane, sorghum and bajra but not in the other three crops such as cotton, groundnut and pulses in irrigated farm. In rainfed farm, animate energy was significant only for cotton and pulses. But the variable has positive influence in pulses, though marginal and a negative influence on cotton.

(3) Mechanical energy use has positive influence on paddy and cotton. The level of significance is at p=0.001 in paddy and p=0.01 in cotton. The level of significance is high in paddy, because it requires more water compared to other crops. The mechanical energy use is near proportionate to the area in dry farms.

(4) In the irrigated lands chemical energy use has positive influence on groundnut only. For the remaining crops, the above energy use was proportionate with cropped area.

(5) In the operation-wise energy consumption regression co-efficient models, the bio-energy had a positive influence with p=0.001 level of significance in all irrigated farms but not much in the dry farm.

(6) Among the various commercial energies used, mechanical energy had a positive influence on large farms of dry land, whereas it had a negative influence on marginal farms of irrigated land. This is due to maximum or surplus use of mechanical energy in the irrigated lands.

7.4 Recommendations

The following major recommendations can be made on the basis of the above analysis.

(1) In order to gain the real picture of energy consumption pattern of various region, number of micro-level studies should be carried out thereby formulate appropriate energy plan for that regions.

(2) Most of these studies are concentrated in the crop-wise energy analysis alone. If necessary a system of pricing structure/taxes has to be evolved to steer up the production pattern in a desired direction.

(3) The use of nitrogen as a fertiliser accounts for a major share of indirect energy consumption by various crops in the study area. Methods should be devised to increase the nitrogen fixing capabilities of the soil either through an appropriate use of crop rotation or through inter-cropping.

(4) The increasing pressure on land will demand much higher yields particularly in the case of food crops such as paddy, sorghum and bajra, which are staple crops in study area as well as in the other parts of India. As seen from the data in previous chapters, paddy, sugarcane, irrigated sorghum and bajra are seen to be the most energy intensive crops. Agro-chemicals and irrigation account for a large part of direct energy consumption. Research efforts should be directed towards increasing the yield of these and other crops without necessarily increasing the input requirements. Efforts should also be made towards improving the quality of low energy intensive cereals such as dry land sorghum, bajra, etc.

7.5 Suggested Steps for Better Energy Management in Agriculture

Issue is quite complex not only technologically but also socio-economically. However, the following points can be considered for better management of energy in agriculture.

(1) Continued monitoring of direct energy inputs to agriculture production, post-harvesting industries, and rural living, identifying energy wasteful operation and their magnitude.

(2) Intensive R&D efforts to have energy efficient cropping system equipment and farm machinery management that assure higher productivity with reduced energy inputs.

(3) Integration of renewable energy sources into overall energy management.

(4) At least 15-20 per cent of diesel and electricity can be saved by having tractor with specific fuel consumption of 214-228 g/kWh instead of 268-295 g/kWh. The implements used must be with BIS (Bureau of Indian Standards) marking.

(5) Transfer of proven technologies to farmers and rural housewives in respect to energy conservation and use of renewable energy through energy management extension wing created in department of agriculture, horticulture and agricultural engineering.

(6) Necessary economic incentives should be given to subsidise energy inputs to small and marginal farmers.

(7) Employing of agricultural engineers trained in energy management for harnessing and explication of renewable sources of energy.

(8) Using of trained techniques in adopting energy efficient devices.

(9) Training the rural women and men in conserving energy resources and optimum utilisation of the scarce resources.

(10) Adequate financial assistance by the banks for renewable energy utilising devices.

(11) Fast growing trees could be planted in the wasteland of the village by implementing the social forestry, farm forestry and agro-forestry programmes.

(12) Rural development programmes could be integrated with rural energy programmes.

(13) Knowledge about the renewable energy sources and rural energy system could be disseminated through distribution of pamphlets, leaflets, film shows, radio and T.V. programmes.

7.6 Alternative Policies

(1) Supporting and extending alternative agricultural practices (perma-culture, natural farming, LEISA, Bio-dynamic etc.)

(2) Promoting diversified farming, integrating with livestock management for greater economic viability and economic stability. De-emphasising on mono-cultures (in poultry, dairy, plants).

(3) Controlling and limiting of export of agriculture produce (especially of grains) for greater food security and sustainability.

(4) Supporting conservation practices through incentives and subsidies to chisel ploughing, check dams, contour bunting, percolation ponds etc. The subsidies should be extended to organic manure, bio-fertilisers, and bio-pesticides instead of chemical fertilisers. This policy may change the attitude of the farmers towards bio-energy use owing to high price of chemical fertilisers.

(5) Specific seed policy would be adopted to improve traditional varieties using modern scientific techniques and knowledge (to stop import and proliferation of exotic seeds, package deals of multi-nationals). The traditional vigour varieties can tolerate any disease.

(6) Imposing compulsory Environment Impact Assessment (EIA)/ Territory Assessment (TA) before introduction of any new technology or seeds.

(7) Land should be distributed to different uses in such a way that they do not affect the quality of the land and into sustainability.

(8) Water resources especially in drought prone areas should be used economically to avoid acute drought and promote dry land farming.

(9) Limiting mechanisation and improving traditional cattle breeds, implements and techniques.

(10) Incentives for green manure, composting and other slurries, limiting import of raw materials for production chemical inputs for agriculture production.

(11) Returning social harness systems in respect of consumables, fuel etc.

(12) Emphasising agriculture as component of environment at school or college level education and extension education given by government departments and voluntary agencies.

(13) Re-examining ownership patterns privatisation, corporatisation), the priorities for community based ownership for sustaining poorer sections of the society.

(14) Preventing or controlling industrialisation and urbanisation - especially in fertile lands.

(15) Promoting decentralised industries supporting agriculture.

7.7 Conclusion

A perusal of the various findings of this study leads to a number of important conclusions. Since there is considerable inter-village and intra-village variation in the energy consumption pattern of this region, it is imperative to have a comprehensive micro-level energy planning for the maximum utilisation of the energy resources in general and non-commercial energy resources in particular. Naturally such a plan would lead to enhanced production resulting ultimately in balanced and sustainable development of the region. In formulating such a plan, would also be helpful to the planners, researchers, energy scientists, policy-makers etc. Furthermore, this study has brought to the limelight the fact that the assured irrigation has played greater role in the level of energy inputs consumption as well as efficient uses of resources. Both village-wise and crop-wise analysis has proved the same. Inadequate irrigation leads to inefficient use of energy resources.

Apart from these, the study has also highlighted the importance of Integrated Energy Management (IEM) approach for the efficient use of energy resources and improving the productivity with these resources. An appropriate combination of non-commercial energy, commercial and bio-energy has proved to be more efficient energy consumption pattern than the haphazard use of these energy resources. It is all the more interesting to note that all the developing countries are facing the paradoxical problem of increasing food production without degrading the agro-ecosystem and environment from over utilisation of chemical energies. As this study has highlighted, the appropriate combination of the energy resources is a panacea to get rid of these kind of complicated problems.

Moreover, as a corollary of this study, it is established that appropriate cropping pattern would serve to achieve self-sufficiency in the energy consumption through higher fixation of nitrogen, improve soil fertility by increasing human content etc. Generally, bio-energy is more promising in terms of its contribution of positive output change. Therefore, extension services to help farmers to switch over to bio-energy rather than commercial energy (chemical fertilisers and pesticides) will go a long way in achieving eco-friendly sustainable development in agriculture.

7.8 Limitations of the Study

This study analyses the issues with respect to the crop year 1997-98. Though its main focus is on the agriculture sector, it takes into consideration that the vital linkages of this sector with other sectors in the rural areas. This present study is basically exploratory in character. It compares with irrigated agriculture, which has its own limitations. The study has to consider the effect of uncontrollable inputs like weather and soil type. Here, these have been assumed to be same on all sample farms. Though the types of soil are different on the farms considered, the parameters of soil like texture and nutrient content were not differentiated. The study has been based mainly on the primary data obtained from farmers. In the absence of published records on income disparity and other information, it was finalised only on the basis of general village records and in consultation with elderly people, educated people and revenue officials of the village.

7.9 Areas of Further Research

The impact of energy input shortage (commercial energy like fertilisers and pesticides) and its subsidies on agriculture production is an important aspect requiring attention. A study on the positive as well as negative effect of the subsidy policy on chemical fertilisers in different states is a potent area for further research. Another area that can be considered for future research, which will be useful in finding out the sustainable future technologies, is the study of the economic and environment viability of renewable energy technology. An analysis of the operation of input energy markets and the relationship between the prices of marketed bio-energy and fossil energy or chemical energy will be a potent area for further research. This will be useful for framing energy subsidisation policies. The following are the need for further research:

1. To optimise animal-machine systems for agricultural operations.

2. To design, develop and evaluate animal transport systems for different regions.

3. To develop and evaluate animal-draft application system for operations like inter-cultural harvesting, auxiliary engine assisted operations like rotary tilling development of efficient mechanical system for conservation and use of animal power for high speed rotary power applications, like water lifting, threshing, winnowing, etc.

References

ABE: (1995) 'Towards a Perspective on Energy Demand and Supply in India in 2004-05', **Advisory Board on Energy,** Govt. of India.

ABERNETHY, V. (1993) 'The Demographic Transition Revisited Lessons for Foreign Aid and U.S. Irrigation Policy', **Ecological Economics,** p. 69.

AICRP: (1980) '**Seventh Annual Report** of AICRP on Energy Requirements in Agricultural Sector (1978-79)', Department of Farm Machinery, **TNAU**, Coimbatore.

ANDROSKY, A: (1973) 'Uses of the Sun in the Service of Man', EI Segundo, California, **Aerospace Corporation Report,** No. ATR 74 (9470), p. 35.

ANWAR ALAM: (1994) 'Energy Perspective and Indian Agriculture, Energy Management and Conservation in Agricultural Production and Food Processing', **USG Publishers and Distributors**, Ludhiana, pp. 59-62.

AULANI, P.K. and W.J.CHANCELLOR, (1977), 'Energy Requirements For Wheat Production and Use in California', **Trans. Am. Soc. Agric. Engineers**, 20, pp. 429-437.

BACHER, KEN: (1974) 'Putting The Sun To Work', Phoenise, Arizona, **Arizona State Fuel and Energy Office**, p. 27.

BANSAL, N.K. (1988), 'Energy Perspective in Rural Development', **Urja**, Vol.23, No.3. March 1988, pp.173, 174.

BASSHAM, J.A: (1984) 'Photosynthetic Plants As Renewable Sources of Chemicals and Energy'; **Non-Conventional Energy** (Ed.,), G.Furlan Et.Al, Plenum Press, New York, pp. 510.

BATTY, J.C. HANED S.M. and KELLER J: (1975) 'Energy Inputs To Irrigation', **Journal of Irrigation**, Drain, Divi, ASCE 151 (IRA), pp. 293-307.

BERRY, R.S. and MAKINO, H. (1974). 'Energy Thrift in Packaging and Marketing', **Technological Review**. Vol.76, No.4, p.1.

BHAGAVAN, M.R. and GIRIAPPA, S (1987) 'Class Character of Rural Energy Crisis - Case of Karnataka', **Economic and Political Weekly**, Review of Agriculture, Vol. XXII, No.26, p. A.27.

BHATNAGAR, A.P.: (1993) 'Energy Management in Indian Agriculture - Retrospect and Prospect', Background Papers For First International Workshop on IREP, Vol.2, **Rural Energy Division, Govt. of India**, and MGIIREPD, Bakoli, Delhi-36, April, pp.13-15.

BISALIAH, S: (1985) **'Energy Use in Agriculture and Changing Forestry, Changing Pattern of Energy Use'** (Ed.) Palwinder Singh, Anmol Publication, New Delhi (1992), p.163.

BISWAS, B.C. TEWATIA, R.K. and NARESH PRASAD: (1981) 'Rice Production in India - Strategies For Nineties', **Fertiliser News**, Vol. 36, No.12, December 1991, pp. 61-67.

BISWAS, B.S., YADAV, D.S. and MAHESWARI, S: (1987) 'Fertiliser Use in Cropping System', **Fertiliser News**, Vol. 32, No. 3, March 1987, pp. 23- 34.

BLACK, J.N: (1971) 'Energy Relations in Crop Production-A Preliminary Survey', **Annual Applied Biology**, Vol.1. pp. 277 - 278.

BLOWIN, G.M. and DAVIS, C.H. (1975) 'Energy Requirement For The Production and Distribution of Chemical Fertilisers in The US'. Paper Presented at **Southern Regional Educational Board Meeting**, Atlanta, GA, Oct. 2, 1975.

BOSE, R.K. and SRIVASTAVA, L: (1988) 'Agriculture Energy Requirement in India', **Advisory Board on Energy**, TERI, New Delhi, p.1.

BRADY, S: (1945) 'Bio-Energetic and Growth', **Reinhold Publishing Corp.**, New York.

CASTELLI, G. and MAZZETTO, F: (1988) 'Manual on Design and Assessment of Integrated Farm Energy System', Project Co-Ordination, PER/80/001- **Development of Integrated Farm Energy Systems**, FAO-UNDP, European Regional Project, pp.118.

CHAMPAN, P.F: (1974) 'Energy Costs: A Review of Methods', **Energy Policy**, Vol. 2, No. 2, pp. 9 13.

CHAWLA, O.P. (1986) 'Advances in Biogas Technology', Publication and Information Division, **Indian Council of Agricultural Research**, New Delhi.

CHRISTOPHER HURST: (1986) 'Energy For Small Scale Engines: A Review of Potential Energy Sources', **Energy Economics** (Ed.), Asok V.Desai, International Development Research Centre, Ottawa, Wiley Eastern Ltd., p.140.

COUGHENOUR, M.B. and OTHERS: (1985) 'Energy Extraction and Use in A Nomadic Pastoral Ecosystem', **Science**, Vol.230, No.4726, Nov.1985, pp. 619 628.

CRAMBIE ABER, K. (1978) ' Which Energy Crisis', **Ceres,** Vol.II (5). pp.13-18.

CROP PRODUCTION GUIDE (1991) **Directorate of Agriculture and Tamil Nadu Agriculture University**, Coimbatore.

CULLUMBINE, H. (1950) 'Heat Production and Energy Requirements of Tropical People', **Journal of Applied Physiology**, No.2, pp. 640-653.

DANIELS, F: (1964) 'Direct Use of the Sun's Energy', **New Haven**, Conn., Yale University Press, p. 374.

DAVID O.HALL: (1980) 'Biological and Agricultural Systems: An Overview'. **Biochemical and Photosynthetic Aspects of Energy Production**, (Ed.,) A.S.Pietro, Academic Press, New York, p. 3.

DAVIS C.H., and BLOVIN G.M. (1977). 'Energy Consumption in the US. Chemical Fertiliser System from the Ground to the Ground', **Agriculture and Energy** (Ed) Lockeretz, W., Academic Press New York, Sept. 1974.

DAYAL, M. (1983), Energy : 'Today and Tomorrow'. **Publication Division**, Govt. of India, New Delhi, p. 3.

DE JAGAR, A. and VAN DER WERF, E. (1992) 'List of Terms, Ecological Agriculture in South India - An Agro-Economic Comparison and Study of Transition', **ETC-Foundation, Lensden**, The Netherlands, pp.70-71.

DE WIT, C.T. (1979), 'The Efficient Use of Labour, Land and Energy in Agriculture'. **Agricu. Systems**, 4: pp. 287-297.

DEKKERS, W.A., LANGE, J.M. and DE WIT. C.T: (1974) 'Energy Production and Its Use in Dutch Agriculture', Netherlands, **Journal of Agric. Sci.**, Vol.22, pp.107 -118.

DEORING, O.C. (1977), 'An Energy Based Analysis of Alternative Production Methods and Cropping System in the Corn Belt', **Report of the Agricultural Research Station**, Purdue University, West Lafyette, p.5.

DEPT. OF AGRIL. ENGINEERING (1973-74) 'Annual Report of Energy Requirement For Intensive Agril. Production', **G.B.Pant University of Agri. and Technology**, pp. 37-42.

DHANAPAL, D: (1991) 'Energy Use Pattern in Agriculture in A Changing Tribal Area † A Micro Level Analysis', **Urja**, Vol.29, No.2 Feb.1991, pp.79 -86.

DHAVAN, K.C., MITTAL, J.P. and BHULLER, B.S.: (1993) 'Energy Scenario of Indian Agriculture and Its Future Requirement', **Energy**, Vol. 18, No. 6, p. 653.

DHAWAN K.C. and SINGH, A.J, K.C: (1976) 'Energy Requirements at Different Levels of Technology in Pubjab', **Indian Journal of Agricultural Economics**, Vol.31, No. 3, pp. 243- 244.

DHAWAN, B.D: (1988) **Agricultural Inputs, Changing of Pattern of Energy Use** (Ed.) Palwinder Singh (1992), Anmol Publication, New Delhi, p.187.

DHAWAN, K.C. and MITTAL, J.P: (1990) 'An Economic Analysis of Energy Requirements For Cotton Under Different Methods of Irrigation', **Economics Affairs**, Vol.35, Qr.2, June 1990, pp. 81-87.

DHAWAN, K.C. and MITTAL, J.P: (1990) 'Energy Scenario of Paddy and Wheat Production in India', **Journal of Rural Development**, NIRD, Hyderabad, Vol.9, No.4, pp.719-728.

DORF, R.C. (1978) **Energy Resources and Policies**, Addison Wesley Publishing Co., Massachusetts, p. 31.

DVOSKIN, D., et.al. (1977), 'Irrigation Energy Requirement in the 17 Western States', **Agriculture and Energy** (Ed) W.Lockeretz, pp.103-112, Academic Press, New York.

ECONOMIC SURVEY (1991-92) **Government of India**, Ministry of Finance (Economic Division) pp. 24-25.

ECONOMIC SURVEY: (1992-93) Agricultural Production, **Government of India**, Ministry of Finance (Economic Division), p.19.

EDWARD SAOUNA: (1984) **Energy For World Agriculture** (Ed.,) Stout, BA., FAO of United Nations.

EDWARDSON, W., LEWIS, and SLESSER, M: (1977) **Food Policy,** May 1977, pp.123 -129.

ENVIRONMENT, Vol.43, Jan.1993, **Elsevier Science Publishers**, Amsterdam, pp.109 -118.

FAI: (1991) Fertiliser Statistics, **The Fertiliser Association of India**, Shasheedjit Singh Marg, New Delhi, pp.11-28.

FAO: (1981) **Energy for World Agriculture** (Ed.,), B.A. Stout, FAO Agri, Series, No.7, Rome.

FLUCK, R.C. AND BAIRD, D.C: (1980) **Agricultural Energetics**, Connecticut, AVI Publishing Company.

FLUCK, R.C: (1979) 'Energy Productivity: A Measure of Energy Utilisation in Agricultural System', **Agricultural System**, Vol. 4, pp. 29-37.

GERALD, A. LEECH: (1989) 'Residential Energy in the Third World', **Annual Review of Energy**, Vol.13, p. 47.

GIAMPIETRO, M. and PIMENTAL, D.: (1993) 'The Tightening Conflict: Population, Energy Use, and the Ecology of Agriculture', **Negative Population Growth**, The NPG Forum, pp.1-7.

GNANAM, A: (1990) A Note From the Guest Editor, **Biology Education**, Vol.7, No.1. p. 4.

GOE, M.R. and MCDOWELL,R.C. (1980) **Animal Traction: Guidelines for Utilisation**, Cornell University Intl. Mimeo, Ithaca, New York.

GOOD, N.E. and BELL, D.H: (1980) **The Biology of Crop Productivity**, Academic Press New York.

GULATI, A and SHARMA, A.N: (1992) 'Subsiding Agriculture: A Cross-Country View', **Economic and Political Weekly**, Vol. XXXVII, No. 38, September 26.

GUPTA R.S.R: (1984) 'Energy Utilisation in Maize Production', **Energy** Vol. 9, (2) Feb. 84, pp.189 -192.

GUPTA, C.L. and RAO, U.K: (1981) 'Energy Inputs For Irrigated Farming With Mixed Cropping: A Micro Level Survey in Pondicherry Region', **Urja**, Vol. 12, No. 5, May 1981, pp.1 - 6.

GUPTA, R.S.R., SHARMA, A.P. and RAO, A.R: (1987) 'Study on Labour and Power Inputs For Agriculture Production in

Harikot Village of Haryana State, India', **Energy in Agriculture**, Vol. 6, No. 1, May, 1987, pp. 57-61.

GURUDEEP CHHATWAL (1987) **Dictionary of Agriculture**, Anmol Publications, New Delhi, India, pp. 9-121.

HALL, D.O and MOSS, P.A: (1983) 'Biomass For Energy in Developing Countries', **Geo Journal**, Vol.7, No.1, pp. 5 - 14.

HEGDE, D.M. and DWIVEDI, B.S: (1992) 'Nutrient Management in Rice-Wheat Cropping System in India', **Fertiliser News**, Vol. 37, No. 2, pp. 27 -41.

HEICHEL G.H. and C.R.FRANK, (1975), 'Anticipating the Energy Needs of American Agriculture', **Journal of Soil and Water Conservation**, 30, pp. 48 - 53.

HOPPER, G.R: (1984), 'The Energetics of Canadian Prairie Wheat Production 1981 – 84', **Energy Developments** (Ed.,), Curtis, F.A., Pergamon Press, New York, pp. 677 - 688.

IFIAS: (1974) International Federation of Institutes for Advanced Study, **Energy Analysis**, Workshop Report No. 6, IFIAS Stockholm.

IUCN/UNEP/WWP (1991) 'Caring for the Earth - A Strategy For Sustainable Living', **Glaud**, Switzerland.

JOHNSON, A.W: (1977) 'Energy Conservation in Amish', **Agriculture Science**, Vol.198, No. 4315, pp. 373- 378.

JONES, D.W: (1988) **Energy Requirements for Rural Development**, Report No:ORNL. 6468, [Ook Ridge: National Laboratory].

KAMATH, M.G. (1972) **A Dictionary of Indian Agriculture**, First Published, p.154.

KAMBLE, H.G. and PANWAR, J.S. (1986) 'Patterns of Selected Energy Input for Seed Production', **Journal of Agriculture Engineering**, Vol. 23, No. 3. p. 279.

KANDA, M: (1975) 'Efficiency For Solar Energy Utilisation: Crop Productivity and Solar Energy Utilisation in Various Climates in Japan', (Ed.,) Y.Murata, JIBS, **Synthesis**, Vol.11, University of Tokyo Press, Tokyo.

KINNE, I.L. and T.A. MACCLURE, (1977), 'Energy in the Food System', **Food, Agriculture and Nutrition** (Ed.) D.N.Lapedes. Mcgraw Hill, New York, pp. 17-32.

KISHIDA, Y: (1973) 'Agro Meteorological Studies on Utilisation of Radiant Energy Under Cultivated Field Conditions', Introduction and Spectral Radiation Balance of Plant Community [In Japanese, English Summery], **Bull Kyushu Agric Exp. Stn.**, Vol.17, pp.1 -79.

KUNDU, T.R., and VASHIST, D.C.: 'Demand For Intermediate Inputs in Indian Agriculture', **Indian Journal of Agriculture Economics**, Vol. 46, No.2, April-June 1991, pp. 152-158.

LARSON, D.D. and D.D.FRAGMEIER, (1977), 'Energy Requirements For Irrigated Crop Production', Proceedings of the **International Conference on Energy Use Management**, Tucson, pp. 734-750.

LEACH, G: (1975) 'Energy and Food Production', **International Institute of Environment and Development**, Washington, D.C.

MACMILLAN, R.H: (1987) 'Engineering Problems in the Measurement of Drought Animal Performance', **Energy and Development** (Ed.,) Jose Goldenberg Et.Al. World

Resource Institute, Oxford & IBH. Pub. Pvt. Ltd., New Delhi, pp.149-155.

MAHESWARI R.C, SRIVASTAVA P.K., BOHRA C.P, TOMAR S.S. and NENA B.P. (1986) 'Energy Census and Resource Assessment of Village Islamnagar in the District of Bhopal', Electro-Mechanical Engineering Division, **Central Institute of Agricultural Engineering**, Bhopal, Second Edition, pp.18-19.

MAHESWARI R.C., et.al (1986) 'Energy Census and Resource Assessment of Village Islamnagar in the District of Bhopal', **Central Institute of Agricultural Engineering**, Electro-Mechanical Engineering Division, Bhopal, pp.1-3.

MAHIJANI, A. and POOLE, A: (1975) 'Energy and Agriculture in the Third World: A Report To the Energy Policy Project of the Food Foundation', **Cambridge**, Ballinger Publishing Co. Mass., USA, p.12.

MALIK, R.K. and RAO, A.R. (1982) 'Energy Audit of Alternative Fuels For Tubewell Irrigation', **Agricultural Engineers Today**, Vol. 6, No. 3 pp. 7-15.

MANI A. and RENGARAJAN, S: (1982) **Solar Radiation over India**, Allied Publishers Pvt. Ltd., New Delhi, p. 646.

MIKCHELL, R: (1979) **The Analysis of Indian Agro-Ecosystem**, New Delhi, Interprint.

MITTAL, J.P. and DHAWAN, K.C. (1989) 'Energy Parameters for Raising Crops under Various Irrigation Treatments in Indian Agriculture', **Agriculture, Ecosystems and Environment**, No. 25, pp.11-25.

MITTAL, V.K., MITTAL J.P. and DHAWAN, K.C. (1985) 'Research Digest on Energy Requirement in Agricultural

Sector', Co-ordinating Cell, AICRP, **Energy Requirement in, Agricultural Sector (ICAR)**, Punjab Agri. University, Ludhiana, pp.7-8; 224 & 159-163.

MONLIK, T.K. and SHUKLA, P.R.:(1989) 'Frame Work for Integrated Energy Planning with Application in India', **Indian Institute of Management**, Ahmedabad, India, Working Paper, No.792, p. 47.

MUNSON, R.D. and DOLL, J.P: (1959) 'The Economics of Fertiliser Use in Crop Production', **Advances in Agronomy**, Vol.II (Ed.,) Norman, A.G., Academic Press, New York, p.133.

MURATA Y., BAE, B.S. and LEE J.Y: (1968) 'On a Convenient Photosynthetic Apparatus for Attached Leaves and Some Results Obtained by Using It' [In Japanese], **Proc. Crop Science Society**, Japan, Vol. 37, pp. 692- 693.

MUTHUCHELIAN, K. and PALIWAL, K: (1988) 'Photosynthesis as a Source of Energy in the Conversion of Solar Energy Into Bio Mass', **Proc. Bio Energy Society Fifth Convention and Symposium 1988**, 30 31, Oct. 1988, Vadodara, pp. 65-66.

NADKARNI, M.V. and DESHPANDE, R.S: (1982) 'Agricultural Growth, Instability in Productivity and Rainfall - Case of Karnataka', **Economic and Political Weekly**, Vol.XVII, No. 52, December 1982.

NAIDU, K.M.S., SEHAIAH, K. PHELIX EDWARD and SANKARAIAH, G: (1983) 'Role of Commercial Energy in Transforming Agriculture', **Energy Development in India** (Ed.,), V.S.Mahajan, Deep and Deep Publications, pp.109 -120.

NARAYANAMOORTHI, A: (1992) 'Impact of Tractors and Weedicides on Yield and Employment in Sugarcane

Cultivation'. **Journal of Rural Development,** Vol.11, No.3, NIRD, Hyderabad, pp. 351 -362, May 1992.

NCA: (1976) 'Report of the National Committee of Agriculture', Part X - Inputs, **Ministry of Agriculture and Irrigation**, Govt. of India, New Delhi.

NCST: (1978) 'An Outlook for India's Future (2000 AD)', Interim Reports of Futurology, **National Committee on Science and Technology**, Govt. of India, New Delhi, India.

NELSON, C.J: (1988) 'Genetic Association Between Photosynthetic Characteristics and Yield', **Plant Physio. Biochemistry**, Vol. 26, pp. 543 -554.

NEWBURY, P.A.R: (1980) **A Geography of Agriculture**, Estover Publishers, Macdonald and Evans Ltd., p. 26.

NINAN, K.N. and CHANDRASEKHAR, H.: (1993) 'Green Revolution, Dryland Agriculture and Sustainability Insights from India', **Economic and Political Weekly**, Vol.28, No.12, 13, March 20-27, pp. A.2- A.6.

NORMAN, M.J.T: (1978) 'Energy Inputs and Outputs of Subsistence Cropping Systems in the Tropics', **Agro Ecosystems**, Vol. 4, pp. 335 -336.

NSEC: (1983) 'Research and Education in The Field of Renewable Sources of Energy', Position Paper, **National Solar Energy Conservation**, Baroda, pp. 2-6.

ODUM, E.P: (1971) **Fundamentals of Ecology**, Saunders, Philadelphia, Pennsylvania, pp. 43 -46.

ODUM, H.T: (1969) 'Energetics of World Food Production', The President's Science Advisory Committee, **Report of Problems of World Food Supply**, Vol.3, Washington D.C., White House.

ODUM, H.T: (1971) 'Environment, Power and Society', **Wiley Inter-science**, New York.

ODUM, H.T: (1983) **Systems Ecology: An Introduction**, New York, Wiley, p. 644.

OECD, 'Organisation for Economic Co-operation and Development', **Energy R and D : Problems and Prospectives**, Paris, 1975, pp.16-17.

OJHA, T.P.: (1993) 'Energy Use in Agriculture', Background Papers for First International Workshop on IREP Organised by Rural Energy Division, **NPC, Govt. of India, and MGIIREPD**, Bakoli, Delhi, Vol. 2, pp.1-12.

OJHA, T.P: (1993) 'Energy Use in Agriculture', Background Papers for First International Workshop on IREP Vol.2 Organised by Rural Energy Division, **National Planning Commission, Govt. of India & MGIIREPD**, Bakoli, Delhi, April, 1993, p. 4.

PACHAURI, R.K: (1985) 'Allocation of Resources for Renewable Energy Development', Workshop on Non†Conventional Energy Planning and Utilisation, **Tata Energy Research Institute**, New Delhi, Dec. 11-13, 1985.

PACHAURI, R.K: (1993) 'Energy and Development', **Journal of Energy in Southern Africa**, Vol. 4, No.1, February 1994, p.14.

PAINULY, J.P., PARIKH, J.K. and SHAH, D.R.: (1992) 'Rural Energy Systems and Agriculture, Alternative Scenario For Gujarat', **Economic and Political Weekly,** August 22, pp. 1801-1809.

PALANISWAMY, K: (1975) 'Study on Energy Use Pattern in Agriculture in Dindugal Division', M.Sc(Agri) Thesis, Dept. of Agri. Economics, **TNAU**, Coimbatore.

PALIWAL, K., NATARAJAN, K. and GNANAM, A: (1990) 'Photosynthesis and Productivity Relationship', **Biology Education**, Vol. 7. No.1. pp. 21-25.

PALWINDER SINGH: (1992) 'Energy Use in Agriculture and Changing Forestry', **Changing Pattern of Energy Use**, Anmol Publication, New Delhi, pp.159-216.

PARTHASARATHI, M: (1973) 'Agricultural Inputs', **Changing Pattern of Energy Use** (Ed.) Palwinder Singh (1992) Anmol Publication, New Delhi, p.189.

PATHAK, B.S: (1985) **Energy in Agriculture**, Vol. 4, p. 67.

PATHAK, B.S: (1987) 'Engineering and Drought Animal Power', **Energy and Development** (Ed.,) John Golden Berg Et.Al., World Resource Instt., Oxford & IBH Pub. Pvt. Ltd., New Delhi, pp.156-160.

PATHAK, R.S. (1983), 'Energy Constraints to Agricultural Production', **Indian Journal of Agric. Econ**. 38, pp. 530-538.

PERELMAN, M: (1972) 'Farming With Petroleum', **Environment**, Vol.14, No.8, pp. 8-13.

PIMENTAL, D and PIMENTAL, M: (1979) **Food, Energy and Society**, Edward Arnoldd, London, UK., p.165.

PIMENTAL, D. and PIMENTAL, M. (1979 & 1980) **Ford, Energy and Society**, Edward Arnold, London, p.165.

PIMENTAL, D., et al: (1973) 'Food Production and the Energy Crisis', **Science**, Vol.182, p. 442.

PIMENTAL, D.L., HURD, E., BELLOTI, A.L., FORSTER, MJ., OKA, J.N., SHOLES, O.D. and WHITMAN, R.J: (1973) 'Food Production and the Energy Crisis', **Science**, Vol.182. pp. 433 -449.

PIMENTAL, D: (1977) 'Energy Budgets in Agricultural and Natural Ecosystems', **Living System As Energy Converters**, (Ed.), Burne, R., Allan, M.J. and Massai, J.P., Amsterdam, North Holland Publishing, pp. 299 -306.

PLANNING COMMISSION: (1979) 'Report of the Working on Energy Policy', **Planning Commission**, Govt. of India, New Delhi, India.

PRAFUL BIDWAI: (1982) 'Agriculture Energy Inputs (Pesticides and Insecticides)', **Changing Pattern of Energy Use** (Ed.), Palwinder Singh (1992) Anmol Publication, New Delhi, pp.193-194.

PRAMOD DEO, MODAK, S. and SHUKLA, P.R. (1991) **Decentralised Energy Planning, Methodology and Case Studies**, Oxford IBH Publishing Co. Pvt. Ltd., New Delhi, pp. 30-33, 76-95.

PRASAD, R. SHARMA, S.N. and SURENDRA SINGH: (1992) 'Fertiliser Nitrogen Interactions in Crop Production', **Fertiliser News**, Vol. 37, No. 4, pp.75 -83.

QUENOVILLE, M.H. et al. (1951) 'Statistical Studies of Recorded Energy Expenditure of Man', **Technical Communication**, No.17, Common Wealth Bureau of Animal Nutrition, Rowett Institute, Scotland.

RAIKHY, P.S. and SINGH, P: (1990) 'Farm Size and Energy Consumption', **Energy Consumption in India † Patterns and Determinants**, Deep and Deep Publications, F-159, Rojouri, Gardens, New Delhi, pp. 41 -67.

RAJA R. and SOORIA MOORTHI, C.E.: (1995) 'Need For Energy Conservation in Agriculture', **Kisan World,** Vol. 22, No. 2, pp. 22-25.

RAJASEKAR, G. and POYYAMOLI, G: (1993) 'Energy Usage in A Village Agroecosystem, Valayambatti, Tamilnadu', M.Phil, Dissertation, **Salim Ali School of Ecology and Environmental Sciences**, Pondicherry University, Pondicherry.

RAJENDRAN, K. SWAMINATHAN, K.R. and MURUGESAN (1987) 'Energy Use Pattern in Farms of Coimbatore District - A Comparative Study of Irrigated and Rainfed Farms'. **Energy Management**, Vol.11, No. 2, April-June 1987, pp.137-141.

RAJA, R. and SOORIAMURTHY, C.E.: (1995) 'Energy Flow in Sugarcane Cultivation in Tamil Nadu', **Energy Management**, Published by National Productivity Council, New Delhi.

RAMASWAMY, N.S. (1993) 'Draught Animal Power', **Journal of Energy**, Vol. 2, Nos. 2-4, p. 3.

RAMESH, N. (1991) 'Studies on Food-Energy Nexus of Pillayarkuppam (Pondicherry U.T.) Village Eco-System', Master's Thesis, **Salim Ali School of Ecology**, Pondicherry University, India.

RAO and BHAVANISHANKAR, M: (1986) 'Agricultural Inputs', **Changing Pattern of Energy Use** (Ed.) Palwinder Singh (1992), Anmol Publication, New Delhi, p.186.

RAO, C.H.H., RAY, S.K. and SUBBA RAO, K: (1988) **Unstable Agriculture and Droughts**, Vikas Publishers, New Delhi.

RAO, Y.S: (1986) 'Plan of Action For Development of Renewable Sources of Rural Energy in the Asia Pacific Region', **Report of the International Workshop**, Part II, Hyderabad, 14-18 April 1986.

RASMUSSEN, S. and A.H. NIELSON, (1978), 'Energy Analysis and Economics', Paper Presented in the **Second Europian Conference of Agricultural Economists,** Dijon, pp. 1-15.

REDDY, A.K.N., et al (1982) (ASTRA), 'Rural Energy Consumption Patterns - A Field Study', **Bio-Mass,** No. 2, p. 256.

REDDY, A.K.N., SUMITRA, G.D., BALACHANDRA, P. and ANTONETTE D'SA (1990) 'Comparative Cost of Electricity Conservation - Centralised and Decentralised Electricity Generation', **Economic and Political Weekly,** pp. 1201-1216.

REDDY, A.K.N: (1981) 'Indian Village Agricultural Eco Systems Case Study of Ungra Village', Part II: Discussion, **Biomass,** Vol.1, No.1, Sept.1881, pp.77- 88.

REPETTO, R. (1986) **World Enough and Time,** New Haven Conn. Yale University Press, p.15.

REVELLE, R.: (1978) 'Requirements for Energy in the Rural Areas of Developing Countries', **Renewable Energy Resources and Rural Application in the Developing World,** (Ed.), Norman.L.Brown, AAAS Selected Symposia Series, Westview Press, Colorado, pp.12-26.

SAHADEV SINGH and GAJENDRA SINGH (1992). 'Energy Input Versus Crop Relationship for Four Major Crops in Northern India', **Agriculture Mechanisation in Asia, Africa and Latin America (AMA),** Vol. 23, No. 2, pp. 57-62.

SATHISH CHANDRAN, T.R. (1977) 'Outlook For Energy in the Coming Decade', **Commerce,** Vol. 3475, pp. 13-15.

SCHAHCZEKSKI, J.J: (1992) 'Energetics and Traditional Agricultural Systems: A Review', **Agricultural Systems**, Vol.14, pp. 31- 43.

SCHIPPER, L. (1994). 'Energy Efficiency Lessons from the Past and Strategies for the Future' Programme of the World Bank Conference on Development Economics, 1993, Supplement To the **World Bank Economic Review and the World Bank Research Observer**, USA, p. 398.

SHARAN, G., MATHUR, D.P. and VISWANATHAN, M: (1975) **Characterisation of the Process of Mechanisation and Farm Power Requirements**, Published by Indian Institute of Management, Ahmedabad, p. 31.

SHARMA, R.K. and RAMESH BHATIA (1987) 'Energy Needs and Low Income Groups in India', **Energy Pricing Policies in Developing Countries** (Ed.) M.S. Kumar, Regional Office For Asia and The Pacific, p.10.

SHARMA, R.N.: (1992) 'Setting Standards in Agriculture' CAPART Clipping (CPC), **Council For Advancement of People's Action and Rural Technology**, New Delhi, pp.16-17.

SHESADRI, C.V: (1992) 'Bio Energy in India: Potential and Prospects', **Energy For Rural Development**, (Ed.), Bhagawan, M.R., and S.Karekezi, Zed Books Ltd., London.

SHUKLA, D.S. P.K. SHARMA, S.R.YADAV and Y.S.CHAUHAN, (1976) 'Pattern of Energy Use Per Hectare in Different Areas of District Bareilly, Uttar Pradesh', **Indian Journal of Agricultural Economics,** Vol. 31, p. 245.

SHYAM, M and NAGAIAH, E.D: (1982) 'Energy Analysis of Potato Production', **Energy Management**, Vol. 6, No. 6, p.199.

SIMON, A.C: (1975), **Energy Sources**, Pengamon Press, Oxford, p.145.

SINGH, A.J. and MIGLANI, SS: (1978) 'An Economic Analysis of Energy Requirements in Punjab Agriculture', **Indian Journal of Agriculture Economics**, Vol. 31, No. 3, p.166.

SINGH, A.J. and MIGLANI,SS (1976), 'An Economic Analysis of Energy Requirements in Punjab Agriculture', **Indian Journal of Agricultural Economics**, Vol. 31, pp.165-173.

SINGH, A.L. and BANERJEE, S: (1988) 'Pattern of Energy Input and Its Impact on Agricultural Production in Agra Division', **The National Geographical Journal of India**, Vol. 34, No. 2, June 1988, pp.126 -133.

SINGH, G and CHANCELLOR, W: (1975) 'Energy Inputs and Agricultural Production Under Various Regions of Mechanisation in Northern India', **Transactions of ASAE**, Vol. 18, No. 2, pp. 252- 259.

SINGH, G.B., SINGH, K.N. and SINGH, R.P: (1990) **Proceedings of International Workshop on Sustainable Land Use System**, New Delhi, pp. 23 -27.

SINGH, L.R and SINGH, B: (1976) 'Level and Pattern of Energy Consumption in Agriculture in Advanced Area of U.P.', **Indian Journal of Agriculture Economics**, Vol. 31, No. 3, pp.143 - 165.

SINGH, R.M. and GOVIL, B.P: (1968) 'Fertiliser Use in Good Crops', **Madras Agriculture Journal**, Vol. 55, p. 225.

SINGH, S.N. BHATNAGAR, S.P. and SINGH R.C: (1991) 'Role of Draught Animal Power in Rural India', **Urja**, Vol. 30, No. 4 Oct.1991, pp. 49-50.

SINHA, S.K: (1986) 'Energy Balance in Agriculture: The Developing World', **Global Aspects of Food Production**

(Ed.), Swaminathan, M.S. and Sinha S.K. London Tycooly International, pp. 47-56.

SINHA, SANRABH: (1986) 'Energy Appropriate Technology and Rural Reconstruction', **Report of International Workshop**, Part II, Hyderabad, April 14-18, 1986.

SIVAKUMAR, S: (1982) 'Estimation of Elasticity Demand for Energy at Farm Level in Three Different Irrigation Regions in Modakurichi Block of Periyar District', M.Sc, Dissertation, Dept. of Agricultural Economics, **TNAU**, Coimbatore, p. 8.

SLESSER, M: (1973) 'Energy Subsidy as a Criteria in Food Policy Planning', **Journal of Science and Food Agriculture**, Vol. 24, pp.1093-1127.

SMERDON, E.T: (1977) 'The Role of Energy in Food Production A Perspective', **Processings of International Conference on Energy Management**, Vol.1 Pergamon Press, Tucson, Arizona, P.963.

SMIL, V. and OTHERS: (1983) 'Technological Changes and the Energy Costs of U.S. Grain Corn', **Energy in Agriculture**, Vol. 2, pp.177-192.

SMIL, V., NACHMAN, P and THOMAS V.LONG II: (1983) 'Solar Energy Conversion Efficiencies in Grain Corn Farming and Consumption', **Energy Analysis and Agriculture-An Application to U.S. Corn. Production,** West View Press, Boulder, Colorado, pp.169-171.

SMIL, V: (1981) 'China's Agro-Ecosystems', **Agro-Ecosystems**, Vol.7, pp. 27-47.

SMIL, VACLAV and KNOWLAND, WILLIAM, H. (1980) **Energy in Developing World**, Oxford University Press, New York, p.5.

SMITH, J.A. and K.J. FORNSTORM, (1980), 'Energy Requirements of Selected Dryland Wheat Cropping Systems'. **Trans. Am. Soc. Agric. Enginers**, 23: pp. 823-825.

SOMASEKARA, N. (1985) **Rural Energy**, Sterling Publishers Pvt. Ltd., New Delhi, p.74.

SPEDDING C.R.W and J.M.WALSINGHAM, (1975), 'Energy Use in Agricultural Systems', **Span**, 18 (1).

SPEDDING C.R.W and J.M.WALSINGHAM, (1976), 'The Production and Use of in Agricultural', **Journal of Agricultural Economics**, 27, pp.19-30.

SRIVASTAVA, N.S.C: (1984) 'Summer Institute of Energy Management and Rural Sector', **CIAE Bhopal**, June 5 - 28, p. 6.

STANHILL, G: (1981) 'The Egyptian Agroecosystem at the End of the 18th Century, An Analysis Based on the Description, De L' Egypte, **Agro-Ecosystems**, Vol. 6, pp. 305 -314.

STEINHART, J.S. and STEINHART, C.E. (1974), 'Energy Use in U.S. Food System', **Science**, 183, pp. 307-316.

STONT, B.A. MYERS, C.A. HURAND, A. and FAIDLEY, L.W. (1979) 'Energy for World Agriculture', **FAO Agriculture Series**, No.7, Food and Agriculture Organisation, Rome, pp. 250-251.

STOUT B.A: (1981) **Energy for World Agriculture**, Publ. by FAO of United Nations, Rome.

STOUT, B.A: (1982) 'Energy for World Agriculture', **Energy Management and Agriculture**, (Ed.), Robinson, D.W. and Mollin, R.C. Royal Dublin Society, pp.1- 21.

SUGGS, L.W. and SPLINTER, W.E. (1961) 'Effect of Environment on the Allowable Workload of Man', **Trans ASAE**, No.4, pp. 48-51.

SURENDER SINGH., SINGH, M.P. and BAKHSHI, R: (1990) 'Unit Energy Consumption For Paddy-Wheat Rotation', **Energy Conservation and Management**, Vol. 30, No. 2, pp.121 -125.

SWAMINATHAN, M.S: (1977) 'Indian Agriculture at the Crossroads', **Indian Journal of Agriculture Economics**, Vol. 32, No. 32, pp.1 -25.

TANDON, H.L.S: (1991) **Fertiliser Management in Dryland Agriculture**, Fertiliser Development and Consultation Organisation, New Delhi, p. 59.

THAKUR, D.S., YADAVA, R.P. and NADDA, A.C.: (1976) 'Energy Requirement of Commercial Crops in Himachal Pradesh', **Indian Journal of Agri-Economics**, Vol. 31, No. 3, p. 242.

TNAU: (1981) 'Report on the Energy Requirement in Farming Systems', Dept. of Agro Energy', College of Agri. Engineering, **Tamil Nadu Agricultural University**, Coimbatore, p. 98.

TSATSARELIS, T.A.: (1992) 'Energy Flow in Sugar Beet Production in Greece', **American Society of Agricultural Engineers**, Vol. 8 No.5, Sept.1992, pp. 585-589.

VASHEK CERVINKA (1980) 'Fuel and Energy Efficiency', **Handbook of Energy Utilisation in Agriculture** (Ed) Pimental D. CRC Press, Boca, Raton, Fl. p.15.

VEENA, D.R: (1988) '**Rural Energy Consumption: Problems and Prospects** (A Replicable Model for India)', Ashish Publishing House, Punjabi Bagh, New Delhi.

VENKATARAMANI, G: (1994) **Good Year for Foodgrains, Study of Indian Agriculture**, 1994, Published By The Hindu, p.5.

VENKATESVARALU, B and VISPERAS, R.M: (1987) 'Solar Radiation and Rice Productivity', **IRRI Research Paper Series**, No.129. Manila, Philippines, pp.7 -10.

VENU, S. (1980) 'Darkness at Noon', **Eastern Economist**, Vol. 3475, pp. 168-170.

White, J.D. (1976), 'Energy Use in Agriculture', **Aspects of Energy Conservation** (Ed.), I.M.Blaire, pp.1441-1476, Pergamon, Oxford.

WILLIAM LOCKERETZ (1980). 'Energy Inputs For Nitrogen, Phosphorus and Potash Fertilisers'. **Handbook of Energy Utilisation in Agriculture** (Ed.) D. Pimental CRC Press, Boca, Raton, Fl. pp. 23-24.

YANAGIDA, J.F. and CONWAY, R.K. (1981), 'The Effect of Energy Price Increases on U S Livestock Sector', **Canada Journal of Agricultural Economics**, 29, pp. 295-302.

YOSHIDA, S. (1981) **Fundamentals of Crop Science**, International Rice Research Institute, P.O.Box.933, Manila, Phillippines, p. 269.

ZELITCH, I: (1975) 'Improving the Efficiency of Photosynthesis', **Science**, No. 88, pp. 628-633.

ZELITH, I: (1982) 'The Close Relationship Between Net Photosynthesis and Crop Yield', **Bio Science**, Vol. 32. pp.796 -802.

ZUCHETTO, J. and JONSON, A.M: (1979) 'Total Energy Analysis of Gotland's Agriculture: A Northern Temperature Case Study', **Agro-Ecosystems**, Vol. 5, pp. 329 -344.

Appendix-I
Energy Content of the Input and Output used

Input	*Unit*	*Equalences in kWh / unit*
Animate Energy		
a) man	1 hour	0.55
Women	1 hour	0.44
Child (12-14)	1 hour	0.27
b) Bullock (pair) body weight (352-450 kg)	1 hour	2.79
Electricity	1 kWh	0.996
Petroleum : Petrol	1 litre	13.36
: Diesel	1 litre	10.90
: Kerosene	1 litre	9.16
: Gasoline	1 litre	9.5
: LPGas	1 litre	7.2
: Natural Gas	m^3	11.5
: Coal (Hard)	kg	8.4
: Coal (Soft)	kg	8.3
Machinery		
Tractor 48kw	Hour	11.05
Tractor 35kw	Hour	9.20
Tractor 26 kw	Hour	6.20
Other Stationery Machine	Hour	2.80
Propelled Machine	Hour	18.95
Sprayer	Hour	6.60
Pump	Hour	0.66
Irrigation systems	hm	0.55
Electric Motor	Hour	7.96

(Contd.)

(Contd. Appendix-I)

Input	*Unit*	*Equalences in kWh / unit*
Plough:		
Wooden	Hour	2.93
Metallic	Hour	19.96
Fertilizers		
Nitrogen (N) kg	kg	20.61
Phosphate (P_2O_5) kg	kg	3.81
Potassium (K_2O) kg	kg	2.69
Zinc Sulphate (Z_nSO_4)	kg	5.80
Ammonium Nitrate (N) kg	kg	18.00
Ammonium Sulphate (N) kg	kg	22.00
Chilean Nitrate (N) kg	kg	5.80
Chemicals: Herbicides	kg	116.00
MCPA	kg	35.50
Insecticides/Pesticides	kg	101.00
Fungicides	kg	27.50
Superior chemicals	kg	33.25
Inferior chemicals (Not required for dilution for application)	kg	2.77
Others		
Farm Yard Marine	kg	0.84
Green manure	kg	0.84
Gypsum	kg	2.77
Neem Cake	kg	8.11
Cow dung	kg	2.78
PRODUCTS		
Sugarcane	kg	2.70
Sugar lawested mass	kg	1.47
Sugar leaves and taps	kg	4.49
Bagasse (Moisture free)	kg	4.64
Paddy Straw (sun dry)	kg	3.48
Sorghum Stalkes (sun dry)	kg	5.45
Senna Stickey (sun dry)	kg	3.83
Cotton stickes (sun dry)	kg	3.83
Bajra stalkes	kg	5.00

(Contd.)

(Contd. Appendix-I)

Input	*Unit*	*Equalences in kWh / unit*
Groundnut (sun dry)	kg	4.93
Pulses (Black & Green gram)	kg	4.08
Pulses Residue (sun dry0	kg	3.47
Rice Husk	kg	3.48
Rice bran	kg	3.83
Coconut husk	kg	4.64
Leaves common	kg	2.78
Agriculture waste	kg	0.36
Seeds	kg	4.64
Unprocessed seeds	kg	3.53
Paddy (grain)	kg	3.44
Paddy (Stubble)	kg	2.39
Ragi	kg	3.68
Cereals		
Wheat, Maize, Sorghum, Bajra,	kg	4.07
Oilseeds		
Cotton seed, Groundnut, Kernel, Seasamum, rap seed	kg	6.93
Cotton lint	kg	4.64
Groundnut shell	kg	4.73

Energy Input for Transportation :

Bullock cart can carry 1 tonne of payload for 10 km and return to a farm on the same day.

Working days per year	176	10.69 kWh/t/km
Inputs into bullock/day	99.68 kWh	
Input into bullock cart/day	6.70 kWh	
Food inputs to driver/day	3.49 kWh	
Total	106.86 kWh or 10.69 kWh/t/km	

Tractor and Trailer:

Goods are transported 20 km/day and return empty.

Payload 3 tonnes

Tractor and Trailer:

Goods are transported 20 km/day and return empty.

Payload 3 tonnes

Diesel and lubricant (@ 0.22 lit/km)

8.8lit (x13.28 kWh/lit) = 116.90 kWh 2.7 kWh/t/km

Dietary energy for driver	3.49 kWh/day
Fixed input for tractor/day	12.99 kWh
Fixed input for trailer	31.16 kWh
Total	164.54 kWh/3t/20km or 2.7 kWh/t/km

Source: Mittal et.al (1985), Management of farm power, Energy and machinery

Notes

(a) Labour energy components are assumed as food energy of doing a task accessory energy utilised for subsistence and a percentage of energy for social existence.

(b) Energy sequestered in Machinery includes the embodied energy in materials and energy consumed during manufacture. The gross energy required (GRE) is amortized for the life period of the equipment and for average years operating hours.

(c) Large varieties of chemicals are used in the form of pesticides (101 kwh/kg) weedicides (116 kWh/kg) and fungicides (27.5 kWh/kg) etc. Average value is used in the study.

(d) Energy in petrol and diesel which are directly used has its energy components as feed back energy and energy consumed in refining and processing.

(e) Electricity is produced by thermal hydel and nuclear routes. As the thermal constitutes the major components, the equipment has been assumed at as 33.33 kWh/kg of electricity.

References

1. Fluke, R.C (1976) To evaluate labour energy in food production, Agriculture Engineering Vil 57, No.1, pp. 31-32.
2. Pemental D et al (1973) Food Production and the Energy crises Science, Vol. 182, pp. 443, 449.
3. Dahiya A.K. (1985) Integrated Energy Systems for small Farms (Ph.D. Theses) Centre for rural Development and Appropriate Technology, IIT, New Delhi, pp. 31-36.
4. Pattack B.S. et. al (1983), The energy Audit of Crop production System, Research Report, School of Energy Studies Agriculture, Punjab Agriculture University, Ludhiana, India, pp. 46-47.
5. Jarach M (1985), Sui valori di equivalenza per 1 analysis e il bilancia energetic in agricolkura, Rivista di/generia No.2 PP. 102-114 (Reference for Electricity).
6. Pimental D, (1980) Energy inputs for the Production formulation, Packaging and transfer chemical and various pesticides. Hand book of energy utilisation in Agriculture (Ed) Basa Ratan, CLC, pp. 35-42 & 112.

Appendix-II
Tables for Chapter-5, Energy Use Pattern
Table 5.1
Source-wise Energy Requirements in Paddy Production (kWh/acre) in Irrigated Farm

Farm Category	*Human Energy*	*Animal Energy*	*Seed Input*	*FYM*	*Wood Instrument*	*Diesel / Petrol*
Bullock Operated Farm						
MF	618.02	1074.16	64.50	4315.83	418.03	116.85
	6.09	*10.59*	*0.64*	*42.54*	*4.12*	*1.15*
SF	552.60	1018.35	57.33	3730.00	461.08	352.95
	5.98	*11.02*	*0.62*	*40.37*	*4.99*	*3.82*
MeF	528.69	971.85	67.37	3745.00	534.00	100.20
	5.82	*10.71*	*0.74*	*41.25*	*5.88*	*1.10*
LF	548.21	1115.17	71.67	4375.00	571.33	133.68
	5.28	*10.73*	*0.69*	*42.11*	*5.50*	*1.29*
Tractor Operated Farm						
MF	516.46	20.21	64.50	4315.00	0.00	499.18
	5.61	*0.22*	*0.70*	*46.88*	*0.00*	*5.42*
SF	455.02	20.67	57.42	3748.95	0.00	550.99
	5.47	*0.25*	*0.69*	*45.08*	*0.00*	*6.63*
MeF	442.69	20.75	67.37	3745.00	0.00	486.15
	5.43	*0.25*	*0.83*	*45.91*	*0.00*	*5.96*
LF	463.38	20.21	71.67	4350.38	0.00	501.23
	5.03	*0.22*	*0.78*	*47.24*	*0.00*	*5.44*

(Contd.)

(Contd. Table 5.1)

Farm Category	*Fertiliser*	*Pesticide / Insecticide*	*Herbicides*	*Electricity*	*Machinery*	*Total*
Bullock Operated Farm						
MF	1523.46	168.33	145.00	1368.25	332.83	10145.27
	15.02	*1.66*	*1.43*	*13.49*	*3.28*	*100.00*
SF	1312.58	126.25	166.23	1165.50	296.92	9239.80
	14.21	*1.37*	*1.80*	*12.61*	*3.21*	*100.00*
MeF	1654.62	126.25	145.00	971.25	233.50	9077.72
	18.23	*1.39*	*1.60*	*10.70*	*2.57*	*100.00*
LF	1866.29	168.33	181.88	1076.71	280.50	10388.77
	17.96	*1.62*	*1.75*	*10.36*	*2.70*	*100.00*
Tractor Operated Farm						
MF	1523.02	168.33	144.17	1387.50	565.62	9203.99
	16.55	*1.83*	*1.57*	*15.07*	*6.15*	*100.00*
SF	1362.58	126.25	290.00	1165.50	538.18	8315.56
	16.39	*1.52*	*3.49*	*14.02*	*6.47*	*100.00*
MeF	1685.42	126.25	145.00	980.70	458.05	8157.37
	20.66	*1.55*	*1.78*	*12.02*	*5.62*	*100.00*
LF	1866.29	168.33	193.33	1068.38	505.05	9208.24
	20.27	*1.83*	*2.10*	*11.60*	*5.48*	*100.00*

Figures given in *Italics* are percentage to the total
Source: Field Data.

Table 5.2

Operation-wise Energy Requirements in Paddy (kWh/acre) in Irrigated Farm

Farm Category	*Preparatory Cultivation*	*Seed input / Planting*	*Irrigation*	*Interculture Operation*	*Fertiliser Application*
Bullock Operated Farm					
MF	1243.21	97.50	1891.67	73.33	1530.79
	12.25	*0.96*	*18.65*	*0.72*	*15.09*
SF	1054.56	88.50	1589.00	87.50	1369.92
	11.41	*0.96*	*17.20*	*0.95*	*14.83*
MeF	1102.01	100.37	1324.17	301.43	1696.42
	12.14	*1.11*	*14.59*	*3.32*	*18.69*
LF	1248.10	102.17	1456.46	376.55	1880.04
	12.01	*0.98*	*14.02*	*3.62*	*18.10*
Tractor Operated Farm					
MF	241.63	97.50	1891.67	87.50	1530.79
	2.63	*1.06*	*20.55*	*0.95*	*16.63*
SF	227.50	88.50	1589.00	89.58	1369.92
	2.74	*1.06*	*19.11*	*1.08*	*16.47*
MeF	231.16	100.37	1324.17	300.60	1696.42
	2.83	*1.23*	*16.23*	*3.69*	*20.80*
LF	218.87	102.17	1456.46	376.55	1880.04
	2.38	*1.11*	*15.82*	*4.09*	*20.42*

(Contd.)

(Contd. Table 5.2)

Farm Category	Fertiliser Application	Plant Protection	Harvesting	Transport	Manuring	Total
Bullock Operated Farm						
MF	1530.79	282.80	234.83	211.97	4579.17	10145.27
	15.09	*2.79*	*2.31*	*2.09*	*45.14*	*100.00*
SF	1369.92	248.18	227.50	382.77	4191.88	9239.80
	14.83	*2.69*	*2.46*	*4.14*	*45.37*	*100.00*
MeF	1696.42	224.02	211.52	344.84	3772.96	9077.72
	18.69	*2.47*	*2.33*	*3.80*	*41.56*	*100.00*
LF	1880.04	281.89	253.57	398.50	4391.50	10388.77
	18.10	*2.71*	*2.44*	*3.84*	*42.27*	*100.00*
Tractor Operated Farm						
MF	1530.79	282.80	96.28	396.67	4579.17	9203.99
	16.63	*3.07*	*1.05*	*4.31*	*49.75*	*100.00*
SF	1369.92	248.18	90.69	420.32	4191.88	8315.56
	16.47	*2.98*	*1.09*	*5.05*	*50.41*	*100.00*
MeF	1696.42	224.02	125.93	381.75	3772.96	8157.37
	20.80	*2.75*	*1.54*	*4.68*	*46.25*	*100.00*
LF	1880.04	281.89	116.30	384.47	4391.50	9208.24
	20.42	*3.06*	*1.26*	*4.18*	*47.69*	*100.00*

(Contd.)

Table 5.3

Energy Productivity, Intensity and Efficiency in Paddy Cultivation (Irrigated Farm)

Farm Category	*Input*		*Output*		*Energy*	
	(kWh / acr)	*kg / acr (grain+resi)*	*(kWh / acr)*	*Productivity (kg / kWh)*	*Intensity (kWh / kg)*	*Efficiency*
Bullock Operated Farm						
MF	10145.27	2125	7335	0.21	4.77	0.72
SF	9239.80	2142	7394	0.23	4.31	0.80
MeF	9077.72	2094	7230	0.23	4.34	0.80
LF	10388.77	2219	7662	0.21	4.68	0.74
Tractor Operated Farm						
MF	9203.99	2177	7515	0.24	4.23	0.82
SF	8315.56	2121	7322	0.26	3.92	0.88
MeF	8157.37	2025	6991	0.25	4.03	0.86
LF	9208.24	2188	7554	0.24	4.21	0.82

Table 5.4

Source-wise Energy Requirements in Sugarcane Production (kWh/acre) in Irrigated Farm

Farm Category	*Human Energy*	*Animal Energy*	*Seed Input*	*FYM*	*Wood Instrument*	*Diesel / Petrol*
Bullock Operated Farm						
MF	403.33	353.40	6477.56	3500.00	164.33	682.42
	2.82	*2.47*	*45.21*	*24.43*	*1.15*	*4.76*
SF	372.08	409.20	7345.10	3500.00	183.79	810.61
	2.31	*2.54*	*45.55*	*21.70*	*1.14*	*5.03*
MeF	413.42	409.20	9231.05	2800.00	193.33	715.27
	2.42	*2.39*	*54.02*	*16.39*	*1.13*	*4.19*
LF	427.63	409.20	6347.92	3500.00	193.33	753.62
	2.81	*2.69*	*41.66*	*22.97*	*1.27*	*4.95*
Tractor Operated Farm						
MF	370.10	0.00	8079.17	3500.00	0.00	786.24
	2.31	*0.00*	*50.37*	*21.82*	*0.00*	*4.90*
SF	333.67	0.00	7352.35	3500.00	0.00	919.61
	2.13	*0.00*	*46.90*	*22.33*	*0.00*	*5.87*
MeF	374.92	0.00	9233.33	2800.00	0.00	824.27
	2.25	*0.00*	*55.52*	*16.84*	*0.00*	*4.96*
LF	389.13	0.00	6403.62	3500.00	0.00	862.62
	2.63	*0.00*	*43.33*	*23.68*	*0.00*	*5.84*

(Contd.)

(Contd. Table 5.4)

Farm Category	*Fertiliser*	*Pesticide / Insecticide*	*Herbicides*	*Electricity*	*Machinery*	*Total*
Bullock Operated Farm						
MF	2073.92	21.04	96.67	499.50	55.72	14327.89
	14.47	*0.15*	*0.67*	*3.49*	*0.39*	*100.00*
SF	2253.56	21.04	96.67	444.00	689.55	16125.60
	13.98	*0.13*	*0.60*	*2.75*	*4.28*	*100.00*
MeF	2213.25	21.04	96.67	444.00	550.20	17087.43
	12.95	*0.12*	*0.57*	*2.60*	*3.22*	*100.00*
LF	2301.63	21.04	145.00	477.30	659.33	15235.99
	15.11	*0.14*	*0.95*	*3.13*	*4.33*	*100.00*
Tractor Operated Farm						
MF	2073.92	21.04	96.67	499.50	613.92	16040.55
	12.93	*0.13*	*0.60*	*3.11*	*3.83*	*100.00*
SF	2253.56	21.04	96.67	444.00	756.08	15676.98
	14.37	*0.13*	*0.62*	*2.83*	*4.82*	*100.00*
MeF	2218.50	21.04	96.67	444.00	616.53	16629.26
	13.34	*0.13*	*0.58*	*2.67*	*3.71*	*100.00*
LF	2301.63	21.04	96.67	477.30	725.83	14777.83
	15.57	*0.14*	*0.65*	*3.23*	*4.91*	*100.00*

Figures given in *Italics* are percentage to the total

Source: Field Data

Table 5.5
Operation-wise Energy Requirements in Sugarcane (kWh/acre) in Irrigated Farm

Farm Category	*Preparatory Cultivation*	*Seed input/ Planting*	*Irrigation*	*Interculture Operation*	*Fertiliser Application*
Bullock Operated Farm					
MF	469.83	8101.17	681.00	252.77	2140.25
	3.28	*56.54*	*4.75*	*1.76*	*14.94*
SF	556.59	6943.33	605.33	652.10	2322.71
	3.45	*43.06*	*3.75*	*4.04*	*14.40*
MeF	560.13	9255.33	605.33	263.10	2290.65
	3.28	*54.16*	*3.54*	*1.54*	*13.41*
LF	560.13	6369.92	650.73	331.03	2406.53
	3.68	*41.81*	*4.27*	*2.17*	*15.80*
Tractor Operated Farm					
MF	173.56	8101.17	681.00	252.77	2140.25
	1.08	*50.50*	*4.25*	*1.58*	*13.34*
SF	192.03	6943.33	605.33	652.10	2322.71
	1.22	*44.29*	*3.86*	*4.16*	*14.82*
MeF	195.70	9255.33	605.33	263.10	2290.65
	1.18	*55.66*	*3.64*	*1.58*	*13.77*
LF	195.70	6369.92	650.73	331.03	2406.53
	1.32	*43.10*	*4.40*	*2.24*	*16.28*

(Contd.)

(Contd. Table 5.5)

Farm Category	Plant Protection	Harvesting	Transport	Manuring	Total
Bullock Operated Farm					
MF	35.69	110.00	1105.85	1431.33	14327.89
	0.25	*0.77*	*7.72*	*9.99*	*100.00*
SF	33.03	110.00	1387.84	3514.67	16125.60
	0.20	*0.68*	*8.61*	*21.80*	*100.00*
MeF	35.78	110.00	1152.43	2814.67	17087.43
	0.21	*0.64*	*6.74*	*16.47*	*100.00*
LF	35.78	121.00	1246.22	3514.67	15235.99
	0.23	*0.79*	*8.18*	*23.07*	*100.00*
Tractor Operated Farm					
MF	35.69	110.00	1031.45	3514.67	16040.55
	0.22	*0.69*	*6.43*	*21.91*	*100.00*
SF	33.03	110.00	1303.78	3514.67	15676.98
	0.21	*0.70*	*8.32*	*22.42*	*100.00*
MeF	35.78	110.00	1058.70	2814.67	16629.26
	0.22	*0.66*	*6.37*	*16.93*	*100.00*
LF	35.78	121.00	1152.48	3514.67	14777.83
	0.24	*0.82*	*7.80*	*23.78*	*100.00*

Figures given in *Italics* are percentage to the total
Source: Field Data.

Table 5.6

Energy Productivity, Intensity and Efficiency in Sugarcane Cultivation (Irrigated Farm)

Farm Category	*Input* (*kWh/acr*)	*kg/acr* (*Cane+resi*)	*Output* (*kWh/acr*)	*Productivity* (*kg/kWh*)	*Energy Intensity* (*kWh/kg*)	*Efficiency* (*O/I*)
Bullock Operated Farm						
MF	14327.89	35208	36628	2.46	0.41	2.56
SF	16125.60	24917	42201	1.55	0.65	2.62
MeF	17087.43	28708	38220	1.68	0.60	2.24
LF	15235.99	26000	43794	1.71	0.59	2.87
Tractor Operated Farm						
MF	16040.55	25729	37822	1.60	0.62	2.36
SF	15676.98	28519	41923	1.82	0.55	2.67
MeF	16629.26	26888	39526	1.62	0.62	2.38
LF	14777.83	31958	46979	2.16	0.46	3.18

Table 5.7
Source-wise Energy Requirements in Cotton Production (kWh/acre)

Farm Category	*Human Energy*	*Animal Energy*	*Seed Input*	*FYM*	*Wood Instrument*	*Diesel / Petrol*
Bullock Operated Farm (Irrigated)						
MF	680.50	911.40	23.10	3668.75	473.67	78.16
	8.39	*11.24*	*0.28*	*45.26*	*5.84*	*0.96*
SF	752.81	669.60	28.88	3985.00	473.67	89.07
	9.13	*8.12*	*0.35*	*48.34*	*5.75*	*1.08*
MeF	814.31	911.40	40.21	3675.00	473.67	133.60
	9.52	*10.66*	*0.47*	*42.97*	*5.54*	*1.56*
LF	827.98	911.40	40.43	4375.00	516.67	116.90
	9.12	*10.04*	*0.45*	*48.18*	*5.69*	*1.29*
Tractor Operated Farm (Irrigated)						
MF	654.83	0.00	23.10	3668.75	0.00	1667.35
	6.77	*0.00*	*0.24*	*37.90*	*0.00*	*17.23*
SF	727.15	0.00	28.88	3985.00	0.00	1765.70
	7.17	*0.00*	*0.28*	*39.30*	*0.00*	*17.41*
MeF	784.90	0.00	34.65	3675.00	0.00	1845.81
	7.58	*0.00*	*0.33*	*35.49*	*0.00*	*17.83*
LF	775.48	0.00	40.43	4375.00	0.00	1940.73
	6.73	*0.00*	*0.35*	*37.97*	*0.00*	*16.84*

(Contd.)

(Contd. Table 5.7)

Farm Category	*Fertiliser*	*Pesticide / Insecticide*	*Herbicides*	*Electricity*	*Machinery*	*Total*
Bullock Operated Farm (Irrigated)						
MF	1553.58	294.58	0.00	288.60	134.20	8106.54
	19.16	*3.63*	*0.00*	*3.56*	*1.66*	*100.00*
SF	1521.94	315.62	0.00	266.40	140.80	8243.78
	18.46	*3.83*	*0.00*	*3.23*	*1.71*	*100.00*
MeF	1681.19	336.67	48.33	255.30	182.60	8552.28
	19.66	*3.94*	*0.57*	*2.99*	*2.14*	*100.00*
LF	1521.17	294.58	95.23	222.00	159.50	9080.85
	16.75	*3.24*	*1.05*	*2.44*	*1.76*	*100.00*
Tractor Operated Farm (Irrigated)						
MF	1553.58	294.58	0.00	288.60	1528.90	9679.70
	16.05	*3.04*	*0.00*	*2.98*	*15.79*	*100.00*
SF	1521.94	315.63	0.00	266.40	1529.68	10140.36
	15.01	*3.11*	*0.00*	*2.63*	*15.09*	*100.00*
MeF	1681.10	336.67	70.34	255.30	1671.28	10355.05
	16.23	*3.25*	*0.68*	*2.47*	*16.14*	*100.00*
LF	1571.17	494.58	96.67	272.00	1957.34	11523.39
	13.63	*4.29*	*0.84*	*2.36*	*16.99*	*100.00*

(Contd.)

(Contd. Table 5.7)

Farm Category	*Human Energy*	*Animal Energy*	*Seed Input*	*FYM*	*Wood Instrument*	*Diesel / Petrol*
Bullock Operated Farm (Unirrigated)						
MF	508.35	420.83	34.65	2450.00	369.75	43.40
	11.74	*9.72*	*0.80*	*56.58*	*8.54*	*1.00*
SF	528.83	475.00	34.65	2800.33	350.42	44.17
	11.07	*9.94*	*0.73*	*58.61*	*7.33*	*0.92*
MeF	554.17	491.67	37.54	3325.00	428.75	33.95
	10.23	*9.07*	*0.69*	*61.35*	*7.91*	*0.63*
LF	546.38	500.95	28.88	3307.47	425.33	44.53
	9.80	*8.99*	*0.52*	*59.35*	*7.63*	*0.80*
Tractor Operated Farm (Unirrigated)						
MF	457.35	0.00	28.88	2450.00	0.00	450.22
	10.11	*0.00*	*0.64*	*54.18*	*0.00*	*9.96*
SF	483.42	0.00	34.65	2800.00	0.00	418.75
	9.99	*0.00*	*0.72*	*57.87*	*0.00*	*8.65*
MeF	510.00	0.00	37.54	3325.00	0.00	435.60
	9.32	*0.00*	*0.69*	*60.77*	*0.00*	*7.96*
LF	502.83	0.00	26.38	3500.00	0.00	487.35
	8.49	*0.00*	*0.45*	*59.10*	*0.00*	*8.23*

(Contd.)

(Contd. Table 5.7)

Farm Category	*Fertiliser*	*Pesticide/ Insecticide*	*Herbicides*	*Electricity*	*Machinery*	*Total*
Bullock Operated Farm (Unirrigated)						
MF	338.63	126.25	0.00	0.00	38.50	4330.36
	7.82	*2.92*	*0.00*	*0.00*	*0.89*	*100.00*
SF	353.45	147.29	0.00	0.00	44.00	4778.15
	7.40	*3.08*	*0.00*	*0.00*	*0.92*	*100.00*
MeF	383.52	126.25	0.00	0.00	38.50	5419.34
	7.08	*2.33*	*0.00*	*0.00*	*0.71*	*100.00*
LF	410.41	168.33	96.67	0.00	44.00	5572.94
	7.36	*3.02*	*1.73*	*0.00*	*0.79*	*100.00*
Tractor Operated Farm (Unirrigated)						
MF	338.63	126.25	0.00	0.00	670.57	4521.89
	7.49	*2.79*	*0.00*	*0.00*	*14.83*	*100.00*
SF	353.45	147.29	0.00	0.00	601.21	4838.77
	7.30	*3.04*	*0.00*	*0.00*	*12.42*	*100.00*
MeF	383.52	126.25	0.00	0.00	653.93	5471.84
	7.01	*2.31*	*0.00*	*0.00*	*11.95*	*100.00*
LF	410.37	168.33	96.67	0.00	730.13	5922.05
	6.93	*2.84*	*1.63*	*0.00*	*12.33*	*100.00*

Table 5.8
Operation-wise Energy Requirements in Cotton (kWh/acre)

Farm Category	*Preparatory Cultivation*	*Seed input / Planting*	*Irrigation*	*Interculture Operation*	*Fertiliser Application*
Bullock Operated Farm (Irrigated)					
MF	466.17	59.77	393.47	403.33	1586.92
	5.75	*0.74*	*4.85*	*4.98*	*19.58*
SF	466.17	65.54	363.20	467.50	1558.60
	5.65	*0.80*	*4.41*	*5.67*	*18.91*
MeF	473.50	82.32	348.07	561.67	1717.77
	5.54	*0.96*	*4.07*	*6.57*	*20.09*
LF	473.50	82.59	302.67	646.67	1557.83
	5.21	*0.91*	*3.33*	*7.12*	*17.16*
Tractor Operated Farm (Irrigated)					
MF	192.03	59.77	393.47	403.33	1586.92
	1.98	*0.62*	*4.06*	*4.17*	*16.39*
SF	170.09	65.54	363.20	467.50	1558.60
	1.68	*0.65*	*3.58*	*4.61*	*15.37*
MeF	199.37	82.32	348.07	561.67	1717.77
	1.93	*0.79*	*3.36*	*5.42*	*16.59*
LF	199.52	82.59	302.67	646.67	1557.83
	1.73	*0.72*	*2.63*	*5.61*	*13.52*

(Contd.)

(Contd. Table 5.8)

Farm Category	*Plant Protection*	*Harvesting*	*Transport*	*Manuring*	*Total*
Bullock Operated Farm (Irrigated)					
MF	460.52	57.29	995.67	3683.42	8106.54
	5.68	*0.71*	*12.28*	*45.44*	*100.00*
SF	507.36	58.44	753.64	4003.33	8243.78
	6.15	*0.71*	*9.14*	*48.56*	*100.00*
MeF	616.93	64.40	994.30	3693.33	8552.28
	7.21	*0.75*	*11.63*	*43.19*	*100.00*
LF	537.98	74.79	1016.07	4388.75	9080.85
	5.92	*0.82*	*11.19*	*48.33*	*100.00*
Tractor Operated Farm (Irrigated)					
MF	460.52	57.29	2842.96	3683.42	9679.70
	4.76	*0.59*	*29.37*	*38.05*	*100.00*
SF	507.36	58.44	2946.29	4003.33	10140.36
	5.00	*0.58*	*29.06*	*39.48*	*100.00*
MeF	616.93	64.40	3071.19	3693.33	10355.05
	5.96	*0.62*	*29.66*	*35.67*	*100.00*
LF	537.98	74.79	3732.58	4388.75	11523.39
	4.67	*0.65*	*32.39*	*38.09*	*100.00*

(Contd.)

(Contd. Table 5.8)

Farm Category	*Preparatory Cultivation*	*Seed input/ Planting*	*Irrigation*	*Interculture Operation*	*Fertiliser Application*
Bullock Operated Farm (Unirrigated)					
MF	400.25	45.65	0.00	355.21	452.33
	9.24	*1.05*	*0.00*	*8.20*	*10.45*
SF	366.88	47.23	0.00	363.33	516.04
	7.68	*0.99*	*0.00*	*7.60*	*10.80*
MeF	428.54	52.46	0.00	370.83	559.58
	7.91	*0.97*	*0.00*	*6.84*	*10.33*
LF	417.65	40.54	0.00	463.33	452.33
	7.49	*0.73*	*0.00*	*8.31*	*8.12*
Tractor Operated Farm (Unirrigated)					
MF	133.03	45.65	0.00	355.21	911.08
	2.94	*1.01*	*0.00*	*7.86*	*20.15*
SF	155.34	47.23	0.00	363.33	788.17
	3.21	*0.98*	*0.00*	*7.51*	*16.29*
MeF	178.65	52.46	0.00	370.83	861.98
	3.26	*0.96*	*0.00*	*6.78*	*15.75*
LF	166.02	40.54	0.00	466.67	979.38
	2.80	*0.68*	*0.00*	*7.88*	*16.54*

(Contd.)

(Contd. Table 5.8)

Farm Category	*Plant Protection*	*Harvesting*	*Transport*	*Manuring*	*Total*
Bullock Operated Farm (Unirrigated)					
MF	346.88	206.40	46.98	2476.67	4330.36
	8.01	*4.77*	*1.08*	*57.19*	*100.00*
SF	361.97	244.78	50.42	2827.50	4778.15
	7.58	*5.12*	*1.06*	*59.18*	*100.00*
MeF	393.10	208.15	54.17	3352.50	5419.34
	7.25	*3.84*	*1.00*	*61.86*	*100.00*
LF	418.62	206.40	46.98	3527.08	5572.94
	7.51	*3.70*	*0.84*	*63.29*	*100.00*
Tractor Operated Farm (Unirrigated)					
MF	346.88	206.40	46.98	2476.67	4521.89
	7.67	*4.56*	*1.04*	*54.77*	*100.00*
SF	361.99	244.78	50.42	2827.50	4838.77
	7.48	*5.06*	*1.04*	*58.43*	*100.00*
MeF	393.10	208.15	54.17	3352.50	5471.84
	7.18	*3.80*	*0.99*	*61.27*	*100.00*
LF	418.62	265.83	57.92	3527.08	5922.05
	7.07	*4.49*	*0.98*	*59.56*	*100.00*

Table 5.9
Energy Productivity, Intensity and Efficiency in Cotton Cultivation

Farm Category	*Input (kWh/acr)*	*Input Kg/acr (lint+stick)*	*Output (kWh/acr)*	*Productivity (kg/kWh)*	*Energy Intensity (kWh/kg)*	*Efficiency (O/I)*
Bullock Operated Farm (Irrigated)						
MF	8106.54	1625	6955	0.20	4.99	0.86
SF	8243.78	2021	8857	0.25	4.08	1.07
MeF	8552.28	2083	9259	0.24	4.11	1.08
LF	9080.85	2283	10078	0.25	3.98	1.11
Tractor Operated Farm (Irrigated)						
MF	9679.70	1606	6925	0.17	6.03	0.72
SF	10140.36	1898	8321	0.19	5.34	0.82
MeF	10355.05	2125	9228	0.21	4.87	0.89
LF	11523.39	2144	9507	0.19	5.38	0.82

(Contd.)

(Contd. Table 5.9)

Farm Category	*Input (kWh/acr)*	*Kg/acr (lint+stick)*	*Output (kWh/acr)*	*Productivity (kg/kWh)*	*Energy Intensity (kWh/kg)*	*Efficiency (O/I)*
Bullock Operated Farm (Irrigated)						
MF	8106.54	1625	6955	0.20	4.99	0.86
SF	8243.78	2021	8857	0.25	4.08	1.07
MeF	8552.28	2083	9259	0.24	4.11	1.08
LF	9080.85	2283	10078	0.25	3.98	1.11
Tractor Operated Farm (Irrigated)						
MF	9679.70	1606	6925	0.17	6.03	0.72
SF	10140.36	1898	8321	0.19	5.34	0.82
MeF	10355.05	2125	9228	0.21	4.87	0.89
LF	11523.39	2144	9507	0.19	5.38	0.82

Table 5.10
Source-wise Energy Requirements in Groundnut Production (kWh/acre)

Farm Category	*Human Energy*	*Animal Energy*	*Seed Input*	*FYM*	*Wood Instrument*	*Diesel / Petrol*
Bullock Operated Farm (Irrigated)						
MF	305.94	595.20	346.50	3846.50	464.00	50.35
	4.91	*9.55*	*5.56*	*61.75*	*7.45*	*0.81*
SF	304.56	319.17	317.63	3846.25	367.33	33.40
	5.26	*5.52*	*5.49*	*66.48*	*6.35*	*0.58*
MeF	332.98	499.88	346.50	3553.96	374.58	47.60
	5.74	*8.62*	*5.97*	*61.27*	*6.46*	*0.82*
LF	342.38	539.40	317.63	3146.25	406.00	50.10
	6.16	*9.71*	*5.72*	*56.65*	*7.31*	*0.90*
Tractor Operated Farm (Irrigated)						
MF	266.63	0.00	346.50	3983.73	0.00	431.39
	4.53	*0.00*	*5.89*	*67.72*	*0.00*	*7.33*
SF	272.00	0.00	317.63	3839.58	0.00	374.03
	4.83	*0.00*	*5.64*	*68.21*	*0.00*	*6.64*
MeF	295.27	0.00	346.50	3553.96	0.00	390.73
	5.43	*0.00*	*6.37*	*65.33*	*0.00*	*7.18*
LF	303.06	0.00	317.63	3146.25	0.00	390.73
	5.92	*0.00*	*6.21*	*61.48*	*0.00*	*7.63*

(Contd.)

(Contd. Table 5.10)

Farm Category	*Fertiliser*	*Pesticide/ Insecticide*	*Herbicides*	*Electricity*	*Machinery*	*Total*
Bullock Operated Farm (Irrigated)						
MF	304.35	126.25	0.00	124.38	66.00	6229.47
	4.89	*2.03*	*0.00*	*2.00*	*1.06*	*100.00*
SF	301.92	84.17	0.00	141.53	69.30	5785.25
	5.22	*1.45*	*0.00*	*2.45*	*1.20*	*100.00*
MeF	286.31	126.25	0.00	155.40	77.30	5800.76
	4.94	*2.18*	*0.00*	*2.68*	*1.33*	*100.00*
LF	310.67	126.25	96.67	145.69	72.88	5553.90
	5.59	*2.27*	*1.74*	*2.62*	*1.31*	*100.00*
Tractor Operated Farm (Irrigated)						
MF	304.35	126.25	0.00	124.88	298.87	5882.59
	5.17	*2.15*	*0.00*	*2.12*	*5.08*	*100.00*
SF	322.75	84.17	0.00	141.53	277.22	5628.89
	5.73	*1.50*	*0.00*	*2.51*	*4.92*	*100.00*
MeF	286.31	126.25	0.00	155.40	285.22	5439.63
	5.26	*2.32*	*0.00*	*2.86*	*5.24*	*100.00*
LF	310.67	126.25	96.67	145.69	280.79	5117.73
	6.07	*2.47*	*1.89*	*2.85*	*5.49*	*100.00*

(Contd.)

(Contd. Table 5.10)

Farm Category	*Human Energy*	*Animal Energy*	*Seed Input*	*FYM*	*Wood Instrument*	*Diesel / Petrol*
Bullock Operated Farm (Unirrigated)						
MF	246.50	612.71	360.94	3473.17	531.67	16.70
	4.55	*11.30*	*6.66*	*64.08*	*9.81*	*0.31*
SF	256.50	626.67	418.69	3367.42	544.96	41.75
	4.62	*11.28*	*7.54*	*60.61*	*9.81*	*0.75*
MeF	272.42	665.08	392.70	3229.35	546.17	25.05
	5.09	*12.42*	*7.33*	*60.31*	*10.20*	*0.47*
LF	275.04	650.01	433.13	3961.67	554.63	33.40
	4.35	*10.27*	*6.85*	*62.61*	*8.77*	*0.53*
Tractor Operated Farm (Unirrigated)						
MF	210.60	0.00	360.94	3473.17	0.00	314.59
	4.47	*0.00*	*7.67*	*73.77*	*0.00*	*6.68*
SF	218.21	0.00	418.69	3367.42	0.00	373.43
	4.49	*0.00*	*8.61*	*69.26*	*0.00*	*7.68*
MeF	234.20	0.00	392.70	3229.35	0.00	350.96
	5.07	*0.00*	*8.50*	*69.86*	*0.00*	*7.59*
LF	236.58	0.00	433.13	3961.67	0.00	387.65
	4.17	*0.00*	*7.64*	*69.89*	*0.00*	*6.84*

(Contd.)

(Contd. Table 5.10)

Farm Category	*Fertiliser*	*Pesticide/ Insecticide*	*Herbicides*	*Electricity*	*Machinery*	*Total*
Bullock Operated Farm (Unirrigated)						
MF	107.28	42.08	0.00	0.00	29.08	5420.13
	1.98	*0.78*	*0.00*	*0.00*	*0.54*	*100.00*
SF	147.50	105.21	0.00	0.00	46.93	5555.62
	2.66	*1.89*	*0.00*	*0.00*	*0.84*	*100.00*
MeF	127.69	63.13	0.00	0.00	33.00	5354.58
	2.38	*1.18*	*0.00*	*0.00*	*0.62*	*100.00*
LF	199.96	84.17	96.67	0.00	38.50	6327.16
	3.16	*1.33*	*1.53*	*0.00*	*0.61*	*100.00*
Tractor Operated Farm (Unirrigated)						
MF	107.28	42.08	0.00	0.00	199.40	4708.06
	2.28	*0.89*	*0.00*	*0.00*	*4.24*	*100.00*
SF	143.75	105.21	0.00	0.00	235.28	4861.99
	2.96	*2.16*	*0.00*	*0.00*	*4.84*	*100.00*
MeF	127.69	63.14	0.00	0.00	224.28	4622.33
	2.76	*1.37*	*0.00*	*0.00*	*4.85*	*100.00*
LF	199.97	84.17	96.67	0.00	268.31	5668.14
	3.53	*1.48*	*1.71*	*0.00*	*4.73*	*100.00*

Table 5.11
Operation-wise Energy Requirements in Groundnut (kWh/acre)

Farm Category	*Preparatory Cultivation*	*Seed input / Planting*	*Irrigation*	*Interculture Operation*	*Fertiliser Application*
Bullock Operated Farm (Irrigated)					
MF	507.27	368.50	170.25	44.00	322.69
	8.14	*5.92*	*2.73*	*0.71*	*5.18*
SF	390.53	339.63	192.95	47.67	168.46
	6.75	*5.87*	*3.34*	*0.82*	*2.91*
MeF	477.17	370.33	211.87	49.50	305.56
	8.23	*6.38*	*3.65*	*0.85*	*5.27*
LF	510.93	344.44	198.63	148.00	329.92
	9.20	*6.20*	*3.58*	*2.66*	*5.94*
Tractor Operated Farm (Irrigated)					
MF	394.34	368.50	170.25	44.00	322.70
	6.70	*6.26*	*2.89*	*0.75*	*5.49*
SF	170.09	65.54	263.20	367.50	458.60
	3.02	*1.16*	*4.68*	*6.53*	*8.15*
MeF	199.37	82.32	234.32	361.67	317.77
	3.67	*1.51*	*4.31*	*6.65*	*5.84*
LF	199.52	82.59	302.67	346.67	557.83
	3.90	*1.61*	*5.91*	*6.77*	*10.90*

(Contd.)

(Contd. Table 5.11)

Farm Category	Plant Protection	Harvesting	Transport	Manuring	Total
Bullock Operated Farm (Irrigated)					
MF	227.91	104.50	616.10	3868.25	6229.47
	3.66	*1.68*	*9.89*	*62.10*	*100.00*
SF	169.13	106.33	502.30	3868.25	5785.25
	2.92	*1.84*	*8.68*	*66.86*	*100.00*
MeF	236.85	111.83	459.63	3578.02	5800.76
	4.08	*1.93*	*7.92*	*61.68*	*100.00*
LF	234.10	121.00	502.30	3164.58	5553.90
	4.22	*2.18*	*9.04*	*56.98*	*100.00*
Tractor Operated Farm (Irrigated)					
MF	227.88	104.50	382.18	3868.25	5882.59
	3.87	*1.78*	*6.50*	*65.76*	*100.00*
SF	307.36	58.44	934.83	3003.33	5628.89
	5.46	*1.04*	*16.61*	*53.36*	*100.00*
MeF	316.93	64.40	671.19	3191.67	5439.63
	5.83	*1.18*	*12.34*	*58.67*	*100.00*
LF	432.32	74.79	732.58	2388.75	5117.72
	8.45	*1.46*	*14.31*	*46.68*	*100.00*

(Contd.)

(Contd. Table 5.11)

Farm Category	*Preparatory Cultivation*	*Seed input / Planting*	*Irrigation*	*Interculture Operation*	*Fertiliser Application*
Bullock Operated Farm (Unirrigated)					
MF	501.38	373.77	0.00	64.25	120.12
	9.25	*6.90*	*0.00*	*1.19*	*2.22*
SF	511.46	432.12	0.00	65.00	157.50
	9.21	*7.78*	*0.00*	*1.17*	*2.84*
MeF	536.75	407.37	0.00	66.83	143.19
	10.02	*7.61*	*0.00*	*1.25*	*2.67*
LF	519.71	448.13	0.00	163.17	214.30
	8.21	*7.08*	*0.00*	*2.58*	*3.39*
Tractor Operated Farm (Unirrigated)					
MF	177.40	373.77	0.00	64.25	120.12
	3.77	*7.94*	*0.00*	*1.36*	*2.55*
SF	177.83	432.12	0.00	65.00	157.50
	3.66	*8.89*	*0.00*	*1.34*	*3.24*
MeF	199.56	407.37	0.00	66.83	143.19
	4.32	*8.81*	*0.00*	*1.45*	*3.10*
LF	221.29	448.13	0.00	163.17	214.30
	3.90	*7.91*	*0.00*	*2.88*	*3.78*

(Contd.)

(Contd. Table 5.11)

Farm Category	*Plant Protection*	*Harvesting*	*Transport*	*Manuring*	*Total*
Bullock Operated Farm (Unirrigated)					
MF	91.78	91.00	681.75	3496.08	5420.13
	1.69	*1.68*	*12.58*	*64.50*	*100.00*
SF	199.29	95.42	703.67	3391.17	5555.62
	3.59	*1.72*	*12.67*	*61.04*	*100.00*
MeF	131.59	101.33	713.17	3254.35	5354.58
	2.46	*1.89*	*13.32*	*60.78*	*100.00*
LF	166.90	103.88	721.42	3989.67	6327.16
	2.64	*1.64*	*11.40*	*63.06*	*100.00*
Tractor Operated Farm (Unirrigated)					
MF	91.78	91.00	293.66	3496.08	4708.06
	1.95	*1.93*	*6.24*	*74.26*	*100.00*
SF	199.29	95.42	343.67	3391.17	4861.99
	4.10	*1.96*	*7.07*	*69.75*	*100.00*
MeF	131.59	101.33	318.10	3254.35	4622.33
	2.85	*2.19*	*6.88*	*70.41*	*100.00*
LF	166.90	103.88	360.82	3989.67	5668.14
	2.94	*1.83*	*6.37*	*70.39*	*100.00*

Table 5.12
Energy Productivity, Intensity and Efficiency in Groundnut Cultivation

Farm Category	Input (kWh/acr)	Kg/acr (kernel+resi)	Output (kWh/acr)	Productivity (kg/kWh)	Energy Intensity (kWh/kg)	Efficiency (O/I)
Bullock Operated Farm (Irrigated)						
MF	6229.47	938	4801	0.15	6.64	0.77
SF	5785.25	950	4857	0.16	6.09	0.84
MeF	5800.76	958	4890	0.17	6.05	0.84
LF	5553.90	1021	5215	0.18	5.44	0.94
Tractor Operated Farm (Irrigated)						
MF	5882.59	896	4578	0.15	6.57	0.78
SF	5628.89	938	4791	0.17	6.00	0.85
MeF	5439.63	960	4897	0.18	5.66	0.90
LF	5117.73	1000	5098	0.20	5.12	1.00

(Contd.)

(Contd. Table 5.12)

Farm Category	Input (kWh/acr)	Kg/acr (kernel+resi)	Output (kWh/acr)	Productivity (kg/kWh)	Energy Intensity (kWh/kg)	Efficiency (O/I)
Bullock Operated Farm (Unirrigated)						
MF	5420.13	813	4141	0.15	6.67	0.76
SF	5555.62	925	4716	0.17	6.01	0.85
MeF	5354.58	998	5087	0.19	5.37	0.95
LF	6327.16	1050	5354	0.17	6.03	0.85
Tractor Operated Farm (Unirrigated)						
MF	4708.06	827	4208	0.18	5.69	0.89
SF	4861.99	917	4668	0.19	5.30	0.96
MeF	4622.33	958	4880	0.21	4.82	1.06
LF	5668.14	969	4936	0.17	5.85	0.87

Table 5.13
Source-wise Energy Requirements in Sorghum Production (kWh/acre)

Farm Category	*Human Energy*	*Animal Energy*	*Seed Input*	*FYM*	*Wood Instrument*	*Diesel / Petrol*
Bullock Operated Farm (Irrigated)						
MF	315.77	496.61	18.70	3500.00	221.13	25.05
	5.65	*8.89*	*0.33*	*62.65*	*3.96*	*0.45*
SF	353.22	519.53	20.40	3675.00	236.83	33.40
	5.79	*8.52*	*0.33*	*60.24*	*3.88*	*0.55*
MeF	371.25	532.27	17.00	3325.00	253.75	50.17
	6.39	*9.16*	*0.29*	*57.23*	*4.37*	*0.86*
LF	359.04	511.08	17.00	4025.00	248.92	57.06
	5.51	*7.84*	*0.26*	*61.72*	*3.82*	*0.87*
Tractor Operated Farm (Irrigated)						
MF	271.94	0.00	18.70	3500.00	0.00	392.93
	5.00	*0.00*	*0.34*	*64.32*	*0.00*	*7.22*
SF	307.05	0.00	20.40	3675.00	0.00	428.53
	5.14	*0.00*	*0.34*	*61.56*	*0.00*	*7.18*
MeF	314.58	0.00	17.00	3325.00	0.00	499.79
	5.52	*0.00*	*0.30*	*58.37*	*0.00*	*8.77*
LF	310.08	0.00	17.00	4025.00	0.00	486.25
	4.85	*0.00*	*0.27*	*62.95*	*0.00*	*7.61*

(Contd.)

(Contd. Table 5.13)

Farm Category	*Fertiliser*	*Pesticide/ Insecticide*	*Herbicides*	*Electricity*	*Machinery*	*Total*
Bullock Operated Farm (Irrigated)						
MF	573.95	68.93	0.00	293.38	72.93	5586.44
	10.27	*1.23*	*0.00*	*5.25*	*1.31*	*100.00*
SF	781.04	92.10	0.00	313.00	76.00	6100.53
	12.80	*1.51*	*0.00*	*5.13*	*1.25*	*100.00*
MeF	748.96	126.25	0.00	295.26	89.83	5809.73
	12.89	*2.17*	*0.00*	*5.08*	*1.55*	*100.00*
LF	758.73	124.08	24.17	305.58	91.10	6521.75
	11.63	*1.90*	*0.37*	*4.69*	*1.40*	*100.00*
Tractor Operated Farm (Irrigated)						
MF	573.95	93.47	0.00	293.38	297.48	5441.84
	10.55	*1.72*	*0.00*	*5.39*	*5.47*	*100.00*
SF	781.04	127.62	0.00	313.00	317.19	5969.83
	13.08	*2.14*	*0.00*	*5.24*	*5.31*	*100.00*
MeF	748.96	130.72	0.00	296.33	364.28	5696.67
	13.15	*2.29*	*0.00*	*5.20*	*6.39*	*100.00*
LF	758.73	116.26	24.17	305.59	350.57	6393.64
	11.87	*1.82*	*0.38*	*4.78*	*5.48*	*100.00*

(Contd.)

(Contd. Table 5.13)

Farm Category	Human Energy	Animal Energy	Seed Input	FYM	Wood Instrument	Diesel / Petrol
Bullock Operated Farm (Unirrigated)						
MF	278.30	530.10	27.20	3150.00	207.83	27.84
	6.17	*11.76*	*0.60*	*69.85*	*4.61*	*0.62*
SF	311.67	474.30	28.90	2975.00	204.01	50.10
	7.05	*10.73*	*0.65*	*67.33*	*4.62*	*1.13*
MeF	273.17	567.30	30.60	3325.00	217.50	50.10
	5.49	*11.40*	*0.61*	*66.82*	*4.37*	*1.01*
LF	320.83	595.20	25.50	4025.00	227.17	66.80
	5.48	*10.17*	*0.44*	*68.78*	*3.88*	*1.14*
Tractor Operated Farm (Unirrigated)						
MF	244.15	0.00	27.20	3150.00	0.00	368.46
	5.63	*0.00*	*0.63*	*72.69*	*0.00*	*8.50*
SF	278.23	0.00	28.90	2975.00	0.00	435.01
	6.49	*0.00*	*0.67*	*69.37*	*0.00*	*10.14*
MeF	235.58	0.00	30.60	3325.00	0.00	467.72
	4.91	*0.00*	*0.64*	*69.33*	*0.00*	*9.75*
LF	284.31	0.00	13.00	4150.00	0.00	447.39
	4.94	*0.00*	*0.23*	*72.10*	*0.00*	*7.77*

(Contd.)

(Contd. Table 5.13)

Farm Category	Fertiliser	Pesticide/ Insecticide	Herbicides	Electricity	Machinery	Total
Bullock Operated Farm (Unirrigated)						
MF	164.08	84.17	0.00	0.00	39.88	4509.40
	3.64	*1.87*	*0.00*	*0.00*	*0.88*	*100.00*
SF	198.70	126.25	0.00	0.00	49.50	4418.43
	4.50	*2.86*	*0.00*	*0.00*	*1.12*	*100.00*
MeF	239.73	126.25	96.67	0.00	49.50	4975.81
	4.82	*2.54*	*1.94*	*0.00*	*0.99*	*100.00*
LF	271.27	168.33	96.67	0.00	55.00	5851.77
	4.64	*2.88*	*1.65*	*0.00*	*0.94*	*100.00*
Tractor Operated Farm (Unirrigated)						
MF	211.64	84.17	0.00	0.00	247.79	4333.41
	4.88	*1.94*	*0.00*	*0.00*	*5.72*	*100.00*
SF	198.71	126.25	0.00	0.00	246.60	4288.70
	4.63	*2.94*	*0.00*	*0.00*	*5.75*	*100.00*
MeF	239.73	126.25	96.67	0.00	274.04	4795.59
	5.00	*2.63*	*2.02*	*0.00*	*5.71*	*100.00*
LF	271.27	168.33	96.67	0.00	325.29	5756.26
	4.71	*2.92*	*1.68*	*0.00*	*5.65*	*100.00*

Table 5.14
Operation-wise Energy Requirements in Sorghum (kWh/acre)

Farm Category	*Preparatory Cultivation*	*Seed input/ Planting*	*Irrigation*	*Interculture Operation*	*Fertiliser Application*
Bullock Operated Farm (Irrigated)					
MF	391.73	44.33	363.80	55.00	584.27
	7.01	*0.79*	*6.51*	*0.98*	*10.46*
SF	438.57	47.48	390.82	66.67	792.18
	7.19	*0.78*	*6.41*	*1.09*	*12.99*
MeF	476.18	44.92	369.75	74.17	759.58
	8.20	*0.77*	*6.36*	*1.28*	*13.07*
LF	447.96	43.67	381.56	115.22	769.15
	6.87	*0.67*	*5.85*	*1.77*	*11.79*
Tractor Operated Farm (Irrigated)					
MF	206.33	44.33	363.80	55.00	584.27
	3.79	*0.81*	*6.69*	*1.01*	*10.74*
SF	233.29	47.48	390.82	66.67	792.18
	3.91	*0.80*	*6.55*	*1.12*	*13.27*
MeF	253.07	44.92	369.75	74.17	759.58
	4.44	*0.79*	*6.49*	*1.30*	*13.33*
LF	186.15	43.67	381.56	115.22	769.15
	2.91	*0.68*	*5.97*	*1.80*	*12.03*

(Contd.)

(Contd. Table 5.14)

Farm Category	*Plant Protection*	*Harvesting*	*Transport*	*Manuring*	*Total*
Bullock Operated Farm (Irrigated)					
MF	133.55	240.50	254.92	3518.33	5586.44
	2.39	*4.31*	*4.56*	*62.98*	*100.00*
SF	175.53	242.79	249.21	3697.29	6100.53
	2.88	*3.98*	*4.09*	*60.61*	*100.00*
MeF	238.42	252.71	246.08	3347.92	5809.73
	4.10	*4.35*	*4.24*	*57.63*	*100.00*
LF	214.57	226.00	274.25	4049.38	6521.75
	3.29	*3.47*	*4.21*	*62.09*	*100.00*
Tractor Operated Farm (Irrigated)					
MF	133.55	137.91	398.32	3518.33	5441.84
	2.45	*2.53*	*7.32*	*64.65*	*100.00*
SF	175.53	145.74	420.84	3697.29	5969.83
	2.94	*2.44*	*7.05*	*61.93*	*100.00*
MeF	238.42	154.20	454.65	3347.92	5696.67
	4.19	*2.71*	*7.98*	*58.77*	*100.00*
LF	214.57	146.45	487.50	4049.38	6393.64
	3.36	*2.29*	*7.62*	*63.33*	*100.00*

(Contd.)

(Contd. Table 5.14)

Farm Category	*Preparatory Cultivation*	*Seed input / Planting*	*Irrigation*	*Interculture Operation*	*Fertiliser Application*
Bullock Operated Farm (Unirrigetad)					
MF	462.50	32.70	0.00	82.50	172.33
	10.26	*0.73*	*0.00*	*1.83*	*3.82*
SF	349.67	34.40	0.00	71.50	202.38
	7.91	*0.78*	*0.00*	*1.62*	*4.58*
MeF	496.27	37.93	0.00	140.67	243.40
	9.97	*0.76*	*0.00*	*2.83*	*4.89*
LF	477.17	31.00	0.00	162.67	282.27
	8.15	*0.53*	*0.00*	*2.78*	*4.82*
Tractor Operated Farm (Unirrigated)					
MF	210.81	32.70	0.00	82.50	172.33
	4.86	*0.75*	*0.00*	*1.90*	*3.98*
SF	180.86	34.40	0.00	71.50	202.38
	4.22	*0.80*	*0.00*	*1.67*	*4.72*
MeF	237.00	37.93	0.00	140.67	243.40
	4.94	*0.79*	*0.00*	*2.93*	*5.08*
LF	258.58	31.00	0.00	162.67	282.27
	4.49	*0.54*	*0.00*	*2.83*	*4.90*

(Contd.)

(Contd. Table 5.14)

Farm Category	*Plant Protection*	*Harvesting*	*Transport*	*Manuring*	*Total*
Bullock Operated Farm (Unirrigetad)					
MF	184.88	215.09	191.07	3168.33	4509.40
	4.10	*4.77*	*4.24*	*70.26*	*100.00*
SF	268.10	238.37	253.35	3000.67	4418.43
	6.07	*5.39*	*5.73*	*67.91*	*100.00*
MeF	258.85	258.80	192.90	3347.00	4975.81
	5.20	*5.20*	*3.88*	*67.27*	*100.00*
LF	326.80	273.60	251.27	4047.00	5851.77
	5.58	*4.68*	*4.29*	*69.16*	*100.00*
Tractor Operated Farm (Unirrigated)					
MF	184.88	123.46	358.40	3168.33	4333.41
	4.27	*2.85*	*8.27*	*73.11*	*100.00*
SF	268.10	146.79	384.01	3000.67	4288.70
	6.25	*3.42*	*8.95*	*69.97*	*100.00*
MeF	258.85	148.57	382.18	3347.00	4795.59
	5.40	*3.10*	*7.97*	*69.79*	*100.00*
LF	326.80	154.07	493.88	4047.00	5756.26
	5.68	*2.68*	*8.58*	*70.31*	*100.00*

Table 5.15
Energy Productivity, Intensity and Efficiency in Sorghum Cultivation

Farm Category	*Input (kWh/acr)*	*Kg/acr (grain+resi)*	*Output (kWh/acr)*	*Productivity (kg/kWh)*	*Energy Intensity (kWh/kg)*	*Efficiency (O/I)*
Bullock Operated Farm (Irrigated)						
MF	5586.44	3833	19608	0.69	1.46	3.51
SF	6100.53	4083	20886	0.67	1.49	3.42
MeF	5800.76	3967	20290	0.68	1.46	3.49
LF	6521.75	4667	23870	0.72	1.40	3.66
Tractor Operated Farm (Irrigated)						
MF	5441.84	4021	17545	0.74	1.35	3.22
SF	5969.83	4292	21880	0.72	1.39	3.67
MeF	5696.67	3808	19506	0.67	1.50	3.42
LF	6393.64	4625	23700	0.72	1.38	3.71

(Contd.)

(Contd. Table 5.15)

Farm Category	Input (kWh/acr)	Kg/acr (grain+resi)	Output (kWh/acr)	Productivity (kg/kWh)	Energy Intensity (kWh/kg)	Efficiency (O/I)
Bullock Operated Farm (Unirrigated)						
MF	4509.40	3667	18755	0.81	1.23	4.16
SF	4418.43	3867	19778	0.88	1.14	4.48
MeF	4975.81	4079	20852	0.82	1.22	4.19
LF	5851.77	4258	21758	0.73	1.37	3.72
Tractor Operated Farm (Unirrigated)						
MF	4333.41	3750	19181	0.87	1.16	4.43
SF	4288.70	3833	19608	0.89	1.12	4.57
MeF	4795.59	4019	20551	0.84	1.19	4.29
LF	5756.26	4313	22023	0.75	1.33	3.83

Table 5.16
Source-wise Energy Requirements in Bajra Production (kWh/acre)

Farm Category	*Human Energy*	*Animal Energy*	*Seed Input*	*FYM*	*Wood Instrument*	*Diesel / Petrol*
Bullock Operated Farm (Irrigated)						
MF	334.58	465.00	6.78	3500.00	183.67	0.00
	6.14	*8.54*	*0.12*	*64.28*	*3.37*	*0.00*
SF	382.25	483.60	6.78	3150.00	183.67	0.00
	7.36	*9.32*	*0.13*	*60.68*	*3.54*	*0.00*
MeF	391.88	539.40	8.48	3325.00	190.62	0.00
	7.18	*9.89*	*0.16*	*60.96*	*3.49*	*0.00*
LF	398.75	595.20	6.78	4200.00	140.33	11.13
	6.23	*9.30*	*0.11*	*65.61*	*2.19*	*0.17*
Tractor Operated Farm (Irrigated)						
MF	314.42	0.00	6.78	3500.00	0.00	327.79
	5.92	*0.00*	*0.13*	*65.92*	*0.00*	*6.17*
SF	366.44	0.00	6.78	3150.00	0.00	375.33
	7.14	*0.00*	*0.13*	*61.37*	*0.00*	*7.31*
MeF	372.40	0.00	8.48	3325.00	0.00	303.39
	7.10	*0.00*	*0.16*	*63.44*	*0.00*	*5.79*
LF	371.48	0.00	6.78	4200.00	0.00	338.85
	5.99	*0.00*	*0.11*	*67.67*	*0.00*	*5.46*

(Contd.)

(Contd. Table 5.16)

Farm Category	*Fertiliser*	*Pesticide / Insecticide*	*Herbicides*	*Electricity*	*Machinery*	*Total*
Bullock Operated Farm (Irrigated)						
MF	471.63	343.75	0.00	116.55	23.10	5445.06
	8.66	*6.31*	*0.00*	*2.14*	*0.42*	*100.00*
SF	411.98	346.95	0.00	188.70	37.41	5191.34
	7.94	*6.68*	*0.00*	*3.63*	*0.72*	*100.00*
MeF	473.55	286.46	0.00	199.80	39.60	5454.78
	8.68	*5.25*	*0.00*	*3.66*	*0.73*	*100.00*
LF	452.25	282.71	24.17	222.00	68.17	6401.49
	7.06	*4.42*	*0.38*	*3.47*	*1.06*	*100.00*
Tractor Operated Farm (Irrigated)						
MF	477.13	343.75	0.00	116.55	222.70	5309.12
	8.99	*6.47*	*0.00*	*2.20*	*4.19*	*100.00*
SF	411.98	355.33	0.00	188.70	278.58	5133.15
	8.03	*6.92*	*0.00*	*3.68*	*5.43*	*100.00*
MeF	473.55	286.46	0.00	199.80	272.47	5241.54
	9.03	*5.47*	*0.00*	*3.81*	*5.20*	*100.00*
LF	463.04	286.45	24.17	222.00	293.97	6206.74
	7.46	*4.62*	*0.39*	*3.58*	*4.74*	*100.00*

(Contd.)

(Contd. Table 5.16)

Farm Category	*Human Energy*	*Animal Energy*	*Seed Input*	*FYM*	*Wood Instrument*	*Diesel / Petrol*
Bullock Operated Farm (Unirrigated)						
MF	294.25	632.40	10.18	2800.00	270.67	0.00
	5.90	*12.68*	*0.20*	*56.14*	*5.43*	*0.00*
SF	276.38	632.21	10.18	3325.00	261.00	0.00
	4.91	*11.23*	*0.18*	*59.04*	*4.63*	*0.00*
MeF	284.63	520.80	10.18	3500.00	185.58	0.00
	5.38	*9.85*	*0.19*	*66.19*	*3.51*	*0.00*
LF	258.27	576.60	8.48	3850.00	213.67	16.70
	4.44	*9.92*	*0.15*	*66.22*	*3.68*	*0.29*
Tractor Operated Farm (Unirrigated)						
MF	264.00	0.00	10.18	2800.00	0.00	381.51
	5.65	*0.00*	*0.22*	*59.98*	*0.00*	*8.17*
SF	256.90	0.00	10.18	3325.00	0.00	395.13
	4.80	*0.00*	*0.19*	*62.09*	*0.00*	*7.38*
MeF	264.69	0.00	10.18	3500.00	0.00	350.46
	5.15	*0.00*	*0.20*	*68.14*	*0.00*	*6.82*
LF	241.06	0.00	8.48	3850.00	0.00	395.13
	4.28	*0.00*	*0.15*	*68.43*	*0.00*	*7.02*

(Contd.)

(Contd. Table 5.16)

Farm Category	*Fertiliser*	*Pesticide / Insecticide*	*Herbicides*	*Electricity*	*Machinery*	*Total*
Bullock Operated Farm (Unirrigated)						
MF	318.08	661.83	0.00	0.00	0.00	4987.41
	6.38	*13.27*	*0.00*	*0.00*	*0.00*	*100.00*
SF	362.98	764.02	0.00	0.00	0.00	5631.76
	6.45	*13.57*	*0.00*	*0.00*	*0.00*	*100.00*
MeF	442.50	343.75	0.00	0.00	0.00	5287.43
	8.37	*6.50*	*0.00*	*0.00*	*0.00*	*100.00*
LF	497.67	343.75	24.17	0.00	24.75	5814.05
	8.56	*5.91*	*0.42*	*0.00*	*0.43*	*100.00*
Tractor Operated Farm (Unirrigated)						
MF	318.08	661.83	0.00	0.00	232.87	4668.47
	6.81	*14.18*	*0.00*	*0.00*	*4.99*	*100.00*
SF	362.98	764.02	0.00	0.00	241.18	5355.38
	6.78	*14.27*	*0.00*	*0.00*	*4.50*	*100.00*
MeF	442.50	343.75	0.00	0.00	224.55	5136.12
	8.62	*6.69*	*0.00*	*0.00*	*4.37*	*100.00*
LF	497.67	343.75	24.17	0.00	265.93	5626.18
	8.85	*6.11*	*0.43*	*0.00*	*4.73*	*100.00*

Table 5.17
Operation-wise Energy Requirements in Bajra (kWh/acre)

Farm Category	*Preparatory Cultivation*	*Seed input / Planting*	*Irrigation*	*Interculture Operation*	*Fertiliser Application*
Bullock Operated Farm (Irrigated)					
MF	383.20	47.12	158.90	58.67	477.13
	7.04	*0.87*	*2.92*	*1:08*	*8.76*
SF	330.33	50.78	257.27	58.67	411.98
	6.36	*0.98*	*4.96*	*1.13*	*7.94*
MeF	390.53	52.48	272.40	58.67	473.55
	7.16	*0.96*	*4.99*	*1.08*	*8.68*
LF	416.93	58.12	302.67	90.30	463.04
	6.51	*0.91*	*4.73*	*1.41*	*7.23*
Tractor Operated Farm (Irrigated)					
MF	199.37	47.12	158.90	58.67	477.13
	3.76	*0.89*	*2.99*	*1.11*	*8.99*
SF	273.21	50.78	257.27	58.67	411.98
	5.32	*0.99*	*5.01*	*1.14*	*8.03*
MeF	223.90	52.48	272.40	58.67	473.55
	4.27	*1.00*	*5.20*	*1.12*	*9.03*
LF	228.87	58.12	302.67	90.30	463.04
	3.69	*0.94*	*4.88*	*1.45*	*7.46*

(Contd.)

(Contd. Table 5.17)

Farm Category	*Plant Protection*	*Harvesting*	*Transport*	*Manuring*	*Total*
Bullock Operated Farm (Irrigated)					
MF	369.42	217.93	207.03	3525.67	5445.06
	6.78	*4.00*	*3.80*	*64.75*	*100.00*
SF	377.21	260.37	265.40	3179.33	5191.34
	7.27	*5.02*	*5.11*	*61.24*	*100.00*
MeF	312.13	277.13	267.23	3350.67	5454.78
	5.72	*5.08*	*4.90*	*61.43*	*100.00*
LF	315.79	258.08	267.23	4229.33	6401.49
	4.93	*4.03*	*4.17*	*66.07*	*100.00*
Tractor Operated Farm (Irrigated)					
MF	369.42	134.07	338.79	3525.67	5309.12
	6.96	*2.53*	*6.38*	*66.41*	*100.00*
SF	377.21	155.49	369.21	3179.33	5133.15
	7.35	*3.03*	*7.19*	*61.94*	*100.00*
MeF	312.13	150.49	347.27	3350.67	5241.54
	5.95	*2.87*	*6.63*	*63.93*	*100.00*
LF	315.79	149.40	369.21	4229.33	6206.74
	5.09	*2.41*	*5.95*	*68.14*	*100.00*

(Contd.)

(Contd. Table 5.17)

Farm Category	*Preparatory Cultivation*	*Seed input / Planting*	*Irrigation*	*Interculture Operation*	*Fertiliser Application*
Bullock Operated Farm (Unirrigated)					
MF	541.80	11.09	0.00	66.00	321.75
	10.86	*0.22*	*0.00*	*1.32*	*6.45*
SF	541.80	11.36	0.00	51.33	368.48
	9.62	*0.20*	*0.00*	*0.91*	*6.54*
MeF	361.20	11.55	0.00	55.00	450.75
	6.83	*0.22*	*0.00*	*1.04*	*8.52*
LF	361.20	10.08	0.00	87.62	512.33
	6.21	*0.17*	*0.00*	*1.51*	*8.81*
Tractor Operated Farm (Unirrigated)					
MF	266.05	11.10	0.00	66.00	321.75
	5.70	*0.24*	*0.00*	*1.41*	*6.89*
SF	243.88	11.55	0.00	51.33	368.48
	4.55	*0.22*	*0.00*	*0.96*	*6.88*
MeF	199.54	11.55	0.00	55.00	450.75
	3.88	*0.22*	*0.00*	*1.07*	*8.78*
LF	225.88	10.08	0.00	87.62	512.33
	4.01	*0.18*	*0.00*	*1.56*	*9.11*

(Contd.)

(Contd. Table 5.17)

Farm Category	*Plant Protection*	*Harvesting*	*Transport*	*Manuring*	*Total*
Bullock Operated Farm (Unirrigated)					
MF	687.50	228.93	301.00	2829.33	4987.41
	13.78	*4.59*	*6.04*	*56.73*	*100.00*
SF	793.35	249.37	265.40	3350.67	5631.76
	14.09	*4.43*	*4.71*	*59.50*	*100.00*
MeF	352.00	260.37	263.57	3533.00	5287.43
	6.66	*4.92*	*4.98*	*66.82*	*100.00*
LF	358.42	254.87	350.20	3879.33	5814.05
	6.16	*4.38*	*6.02*	*66.72*	*100.00*
Tractor Operated Farm (Unirrigated)					
MF	687.50	155.90	330.83	2829.33	4668.47
	14.73	*3.34*	*7.09*	*60.61*	*100.00*
SF	793.35	157.74	378.38	3350.67	5355.38
	14.81	*2.95*	*7.07*	*62.57*	*100.00*
MeF	352.00	168.74	365.55	3533.00	5136.12
	6.85	*3.29*	*7.12*	*68.79*	*100.00*
LF	358.42	163.20	389.32	3879.33	5626.18
	6.37	*2.90*	*6.92*	*68.95*	*100.00*

Table 5.18
Energy Productivity, Intensity and Efficiency in Bajra Cultivation

Farm Category	*Input (kWh/acr)*	*Kg/acr (grain+resi)*	*Output (kWh/acr)*	*Productivity (kg/kWh)*	*Energy Intensity (kWh/kg)*	*Efficiency (O/I)*
Bullock Operated Farm (Irrigated)						
MF	5445.06	2750	12936	0.51	1.98	2.38
SF	5191.34	3063	14383	0.59	1.70	2.77
MeF	5454.78	3146	14780	0.58	1.73	2.71
LF	6401.49	3333	15659	0.52	1.92	2.45
Tractor Operated Farm (Irrigated)						
MF	5309.12	2896	13646	0.55	1.83	2.57
SF	5133.15	2988	14027	0.58	1.72	2.73
MeF	5241.54	3092	14521	0.59	1.70	2.77
LF	6206.74	3425	16090	0.55	1.81	2.59

(Contd.)

(Contd. Table 5.18)

Farm Category	*Input (kWh/acr)*	*Kg/acr (grain+resi)*	*Output (kWh/acr)*	*Productivity (kg/kWh)*	*Energy Intensity (kWh/kg)*	*Efficiency (O/I)*
Bullock Operated Farm (Unirrigated)						
MF	4987.41	1333	6124	0.27	3.74	1.23
SF	5631.76	1438	6606	0.26	3.92	1.17
MeF	5287.43	1229	5623	0.23	4.30	1.06
LF	5824.05	1729	7948	0.30	3.36	1.37
Tractor Operated Farm (Unirrigated)						
MF	4668.47	1296	5956	0.28	3.60	1.28
SF	5355.38	1446	6656	0.27	3.70	1.24
MeF	5136.12	1146	5225	0.22	4.48	1.02
LF	5626.18	1629	7483	0.29	3.45	1.33

Table 5.19
Source-wise Energy Requirements in Pulses Production (kWh/acre)

Farm Category	*Human Energy*	*Animal Energy*	*Seed Input*	*FYM*	*Wood Instrument*	*Diesel / Petrol*
Bullock Operated Farm (Irrigated)						
MF	332.02	346.43	32.30	3846.43	180.04	0.00
	6.49	*6.77*	*0.63*	*75.17*	*3.52*	*0.00*
SF	332.29	290.63	35.70	3090.63	148.63	0.00
	7.72	*6.75*	*0.83*	*71.81*	*3.45*	*0.00*
MeF	316.25	390.60	37.40	3325.00	275.25	0.00
	6.42	*7.92*	*0.76*	*67.46*	*5.58*	*0.00*
LF	328.40	409.17	34.00	3500.00	222.33	0.00
	6.28	*7.82*	*0.65*	*66.92*	*4.25*	*0.00*
Tractor Operated Farm (Irrigated)						
MF	251.63	0.00	32.30	3846.43	0.00	231.63
	5.03	*0.00*	*0.65*	*76.87*	*0.00*	*4.63*
SF	284.85	0.00	35.70	3090.63	0.00	286.13
	6.55	*0.00*	*0.82*	*71.03*	*0.00*	*6.58*
MeF	274.08	0.00	37.41	3325.00	0.00	263.60
	5.84	*0.00*	*0.80*	*70.82*	*0.00*	*5.61*
LF	271.33	0.00	34.00	3500.00	0.00	299.75
	5.32	*0.00*	*0.67*	*68.64*	*0.00*	*5.88*

(Contd.)

(Contd. Table 5.19)

Farm Category	*Fertiliser*	*Pesticide / Insecticide*	*Herbicides*	*Electricity*	*Machinery*	*Total*
Bullock Operated Farm (Irrigated)						
MF	0.00	286.46	0.00	77.70	15.40	5116.77
	0.00	*5.60*	*0.00*	*1.52*	*0.30*	*100.00*
SF	0.00	286.46	0.00	99.63	19.80	4303.76
	0.00	*6.66*	*0.00*	*2.31*	*0.46*	*100.00*
MeF	0.00	343.75	0.00	219.34	21.21	4928.80
	0.00	*6.97*	*0.00*	*4.45*	*0.43*	*100.00*
LF	224.40	343.75	0.00	147.94	19.80	5229.78
	4.29	*6.57*	*0.00*	*2.83*	*0.38*	*100.00*
Tractor Operated Farm (Irrigated)						
MF	0.00	286.46	0.00	147.70	207.50	5003.63
	0.00	*5.73*	*0.00*	*2.95*	*4.15*	*100.00*
SF	0.00	286.46	0.00	127.14	240.19	4351.09
	0.00	*6.58*	*0.00*	*2.92*	*5.52*	*100.00*
MeF	0.00	343.75	0.00	219.34	231.58	4694.77
	0.00	*7.32*	*0.00*	*4.67*	*4.93*	*100.00*
LF	224.40	343.75	0.00	174.02	251.82	5099.07
	4.40	*6.74*	*0.00*	*3.41*	*4.94*	*100.00*

(Contd.)

(Contd. Table 5.19)

Farm Category	*Human Energy*	*Animal Energy*	*Seed Input*	*FYM*	*Wood Instrument*	*Diesel / Petrol*
Bullock Operated Farm (Unirrigated)						
MF	266.29	255.75	44.20	2975.00	181.25	0.00
	6.64	*6.38*	*1.10*	*74.21*	*4.52*	*0.00*
SF	295.17	388.28	44.20	3150.00	201.79	0.00
	6.67	*8.78*	*1.00*	*71.23*	*4.56*	*0.00*
MeF	301.04	441.75	42.50	2975.00	229.44	0.00
	6.95	*10.19*	*0.98*	*68.65*	*5.29*	*0.00*
LF	400.50	349.91	42.50	4166.67	181.25	0.00
	7.22	*6.31*	*0.77*	*75.15*	*3.27*	*0.00*
Tractor Operated Farm (Unirrigated)						
MF	230.44	0.00	44.20	2975.00	0.00	274.79
	5.81	*0.00*	*1.11*	*74.96*	*0.00*	*6.92*
SF	258.40	0.00	44.20	3150.00	0.00	282.94
	6.09	*0.00*	*1.04*	*74.21*	*0.00*	*6.67*
MeF	256.71	0.00	42.50	2975.00	0.00	269.40
	6.35	*0.00*	*1.05*	*73.54*	*0.00*	*6.66*
LF	284.06	0.00	42.50	4200.00	0.00	313.38
	5.21	*0.00*	*0.78*	*77.09*	*0.00*	*5.75*

(Contd.)

(Contd. Table 5.19)

Farm Category	Fertiliser	Pesticide / Insecticide	Herbicides	Electricity	Machinery	Total
Bullock Operated Farm (Unirrigated)						
MF	0.00	286.46	0.00	0.00	0.00	4008.95
	0.00	*7.15*	*0.00*	*0.00*	*0.00*	*100.00*
SF	0.00	342.67	0.00	0.00	0.00	4422.10
	0.00	*7.75*	*0.00*	*0.00*	*0.00*	*100.00*
MeF	0.00	343.75	0.00	0.00	0.00	4333.48
	0.00	*7.93*	*0.00*	*0.00*	*0.00*	*100.00*
LF	117.19	286.46	0.00	0.00	0.00	5544.48
	2.11	*5.17*	*0.00*	*0.00*	*0.00*	*100.00*
Tractor Operated Farm (Unirrigated)						
MF	0.00	286.46	0.00	0.00	158.02	3968.90
	0.00	*7.22*	*0.00*	*0.00*	*3.98*	*100.00*
SF	0.00	342.67	0.00	0.00	166.33	4244.54
	0.00	*8.07*	*0.00*	*0.00*	*3.92*	*100.00*
MeF	0.00	343.75	0.00	0.00	158.01	4045.37
	0.00	*8.50*	*0.00*	*0.00*	*3.91*	*100.00*
LF	117.19	286.46	0.00	0.00	204.45	5448.03
	2.15	*5.26*	*0.00*	*0.00*	*3.75*	*100.00*

(Contd.)

Table 5.20
Operation-wise Energy Requirements in Pulses (kWh/acre)

Farm Category	*Preparatory Cultivation*	*Seed input / Planting*	*Irrigation*	*Interculture Operation*	*Fertiliser Application*
Bullock Operated Farm (Irrigated)					
MF	379.53	34.13	147.60	36.67	0.00
	7.42	*0.67*	*2.88*	*0.72*	*0.00*
SF	388.70	37.53	136.20	51.26	0.00
	9.03	*0.87*	*3.16*	*1.19*	*0.00*
MeF	439.73	39.00	258.16	44.03	0.00
	8.92	*0.79*	*5.24*	*0.89*	*0.00*
LF	563.80	36.06	136.20	47.17	230.73
	10.78	*0.69*	*2.60*	*0.90*	*4.41*
Tractor Operated Farm (Irrigated)					
MF	173.53	34.13	147.60	36.67	0.00
	3.47	*0.68*	*2.95*	*0.73*	*0.00*
SF	227.04	37.53	136.20	51.26	0.00
	5.22	*0.86*	*3.13*	*1.18*	*0.00*
MeF	195.70	39.00	258.16	44.03	0.00
	4.17	*0.83*	*5.50*	*0.94*	*0.00*
LF	221.54	36.06	136.20	47.17	230.73
	4.34	*0.71*	*2.67*	*0.93*	*4.52*

(Contd.)

(Contd. Table 5.20)

Farm Category	*Plant Protection*	*Harvesting*	*Transport*	*Manuring*	*Total*
Bullock Operated Farm (Irrigated)					
MF	301.13	159.50	190.70	3867.51	5116.77
	5.89	*3.12*	*3.73*	*75.58*	*100.00*
SF	304.60	157.67	104.18	3123.63	4303.76
	7.08	*3.66*	*2.42*	*72.58*	*100.00*
MeF	363.00	155.83	276.53	3352.50	4928.80
	7.36	*3.16*	*5.61*	*68.02*	*100.00*
LF	391.42	172.33	130.07	3522.00	5229.78
	7.48	*3.30*	*2.49*	*67.35*	*100.00*
Bullock Operated Farm (Irrigated)					
MF	301.13	159.50	283.56	3867.51	5003.63
	6.02	*3.19*	*5.67*	*77.29*	*100.00*
SF	304.60	157.67	313.17	3123.63	4351.09
	7.00	*3.62*	*7.20*	*71.79*	*100.00*
MeF	363.00	155.83	286.54	3352.50	4694.77
	7.73	*3.32*	*6.10*	*71.41*	*100.00*
LF	391.42	172.33	341.63	3522.00	5099.07
	7.68	*3.38*	*6.70*	*69.07*	*100.00*

(Contd.)

(Contd. Table 5.20)

Farm Category	*Preparatory Cultivation*	*Seed input/ Planting*	*Irrigation*	*Interculture Operation*	*Fertiliser Application*
Bullock Operated Farm (Unirrigated)					
MF	282.87	45.58	0.00	51.33	0.00
	7.06	*1.14*	*0.00*	*1.28*	*0.00*
SF	436.07	46.03	0.00	51.33	0.00
	9.86	*1.04*	*0.00*	*1.16*	*0.00*
MeF	563.80	44.56	0.00	44.00	0.00
	13.01	*1.03*	*0.00*	*1.02*	*0.00*
LF	379.53	44.33	0.00	47.67	130.02
	6.91	*0.81*	*0.00*	*0.87*	*2.37*
Tractor Operated Farm (Unirrigated)					
MF	214.20	45.58	0.00	51.33	0.00
	5.40	*1.15*	*0.00*	*1.29*	*0.00*
SF	213.98	46.03	0.00	51.33	0.00
	5.04	*1.08*	*0.00*	*1.21*	*0.00*
MeF	199.37	44.56	0.00	44.00	0.00
	4.93	*1.10*	*0.00*	*1.09*	*0.00*
LF	217.87	44.33	0.00	47.67	130.02
	4.00	*0.81*	*0.00*	*0.87*	*2.39*

(Contd.)

(Contd. Table 5.20)

Farm Category	*Plant Protection*	*Harvesting*	*Transport*	*Manuring*	*Total*
Bullock Operated Farm (Unirrigated)					
MF	308.46	121.00	195.38	3004.33	4008.95
	7.69	*3.02*	*4.87*	*74.94*	*100.00*
SF	361.00	150.33	201.67	3175.67	4422.10
	8.16	*3.40*	*4.56*	*71.81*	*100.00*
MeF	362.08	146.67	171.70	3000.67	4333.48
	8.36	*3.38*	*3.96*	*69.24*	*100.00*
LF	301.13	166.83	199.30	4225.67	5494.48
	5.48	*3.04*	*3.63*	*76.91*	*100.00*
Tractor Operated Farm (Unirrigated)					
MF	308.46	121.00	224.00	3004.33	3968.90
	7.77	*3.05*	*5.64*	*75.70*	*100.00*
SF	361.00	150.33	246.19	3175.67	4244.54
	8.51	*3.54*	*5.80*	*74.82*	*100.00*
MeF	362.08	146.67	248.03	3000.67	4045.37
	8.95	*3.63*	*6.13*	*74.18*	*100.00*
LF	301.13	166.83	314.52	4225.67	5448.03
	5.53	*3.06*	*5.77*	*77.56*	*100.00*

Table 5.21
Energy Productivity, Intensity and Efficiency in Pulses Cultivation

Farm Category	Input (kWh/acr)	Kg/acr (grain+resi)	Output (kWh/acr)	Productivity (kg/kWh)	Energy Intensity (kWh/kg)	Efficiency (O/I)
Bullock Operated Farm (Irrigated)						
MF	5116.77	746	2830	0.15	6.86	0.55
SF	4303.76	796	3018	0.18	5.41	0.70
MeF	4928.80	867	3287	0.18	5.69	0.67
LF	5229.78	892	3381	0.17	5.87	0.65
Tractor Operated Farm (Irrigated)						
MF	5003.63	746	3805	0.15	6.71	0.76
SF	4351.09	800	2827	0.18	5.44	0.65
MeF	4694.77	867	3030	0.18	5.42	0.65
LF	5099.07	900	3274	0.18	5.67	0.64

(Contd.)

(Contd. Table 5.21)

Farm Category	Input (kWh/acr)	Kg/acr (grain+resi)	Output (kWh/acr)	Productivity (kg/kWh)	Energy Intensity (kWh/kg)	Efficiency (O/I)
Bullock Operated Farm (Unirrigated)						
MF	4008.95	650	2433	0.16	6.17	0.61
SF	4422.10	671	2511	0.15	6.59	0.57
MeF	4333.48	696	2605	0.16	6.23	0.60
LF	5544.48	683	2559	0.12	8.11	0.46
Tractor Operated Farm (Unirrigated)						
MF	3968.90	663	2494	0.17	5.99	0.63
SF	4244.54	683	2557	0.16	6.21	0.60
MeF	4045.37	683	2559	0.17	5.92	0.63
LF	5448.03	660	2473	0.12	8.25	0.45

Appendix -III
Tables for Chapter-6, The Influence of Energy Input

Table 6.1
Village-wise and Source-wise Energy Use Pattern for Agricultural Operations for the year 1997-98 in the Study Area (kWh/annum)

SlNo	*Village*	*Land Used*		*Total*	*Human Energy*	*Animal Energy*	*Seed Input*	*FYM*	*Wood Instrument*
		Irrigated	*Dryland*						
1	Pulkattai	20.55	33.15	53.7	25783.2	30733.2	46157.8	121936.8	13980.2
2	Vannivelampatti	17.9	27.7	45.6	23343.0	39914.2	60461.2	150827.4	22210.4
3	Kunnathur	17.9	28.5	46.4	38369.0	38121.0	55897.8	136552.0	25124.2
4	Velampur	9	26.4	35.4	26552.6	38488.6	71654.2	103111.2	36437.2
5	Genjampatti	9.2	27.1	36.3	28868.0	28952.0	72744.2	161752.6	24405.8
6	Nallamaram	6	30.6	36.6	32277.4	28991.0	39364.6	111150.6	25135.0
7	Silarpatti	10.2	18.2	28.4	36334.0	28897.8	22037.4	66250.2	24867.6
8	Shandaiyur	9.6	31.8	41.4	27803.2	34836.0	25461.2	139838.8	16586.6
9	Melapatti	13.65	33.85	47.5	25122.6	31796.4	30926.6	180465.6	16575.8
10	Pappaiyapuram	10.1	23.85	33.95	29582.2	24259.8	24973.8	113012.6	23659.4
11	Kuvalapuram	7.6	27.1	34.7	31328.8	28531.8	49971.4	96192.0	25836.4
12	Chittulotti	10.5	29.5	40	32036.8	21895.4	54686.4	122142.2	21598.6
	Total	142.2	337.75	479.95	357400.8	375417.2	554336.6	1503232.0	276417.2
	Average				29783.4	31284.8	46194.7	125269.3	23034.8

(Contd.)

(Contd. Table 6.1)

SlNo	Village	Diesel / Petrol	Fertiliser	Pesticide/ Insecticide	Herbicides	Electricity	Machinery	Total
1	Pulkattai	24871.4	34604.2	7572.8	5060.6	34674.6	45039.6	390414.4
2	Vannivelampatti	26091.8	25217.2	16896.2	5316.0	64447.2	77230.2	511954.8
3	Kunnathur	26025.8	36511.4	18159.4	9013.2	56521.6	86801.8	527097.2
4	Velampur	26549.6	24843.2	26251.2	7638.8	45634.0	98895.6	506056.2
5	Genjampatti	24687.8	31021.8	24168.0	5874.8	27591.4	89325.2	519391.6
6	Nallamaram	23051.4	37835.2	25198.4	5733.6	31737.8	73811.6	434286.6
7	Silarpatti	34829.4	30195.0	24808.8	4118.0	53366.0	48417.2	374121.4
8	Shandaiyur	29197.2	29188.4	37122.2	8769.0	29173.4	92884.2	470860.2
9	Melapatti	42443.0	35715.6	20605.0	3032.8	21157.0	73079.8	480920.2
10	Pappaiyapuram	19822.6	25424.8	24219.2	3971.0	33780.8	39138.4	361844.6
11	Kuvalapuram	31042.2	23225.0	25519.8	8099.8	55040.2	85998.8	460786.2
12	Chittulotti	16177.0	27427.2	27439.8	8405.8	23974.6	78665.0	434448.8
	Total	324789.2	361209.0	277960.8	75033.4	477098.6	889287.4	5472182.2
	Average	27065.8	30100.7	23163.4	6252.8	39758.2	74107.3	456015.2

Table 6.2

Village-wise and Source-wise Energy Use for the year 1997-98 in the Study Area (kWh/acre)

SlNo	*Village*	*Human Energy*	*Animal Energy*	*Seed Input*	*FYM*	*Wood Instrument*	*Diesel / Petrol*
1	Pulkattai	480.13	572.31	859.55	2270.70	260.34	463.15
		6.60	*7.87*	*11.82*	*31.23*	*3.58*	*6.37*
2	Vannivelampatti	511.91	875.31	1325.90	3307.62	487.07	572.19
		4.56	*7.80*	*11.81*	*29.46*	*4.34*	*5.10*
3	Kunnathur	826.92	821.57	1204.69	2942.93	541.47	560.90
		7.28	7.23	10.60	25.91	4.77	4.94
4	Velampur	750.07	1087.25	2024.13	2912.75	1029.30	749.99
		5.25	*7.61*	*14.16*	*20.38*	*7.20*	*5.25*
5	Genjampatti	795.26	797.58	2003.97	4455.99	672.34	680.10
		5.56	*5.57*	*14.01*	*31.14*	*4.70*	*4.75*
6	Nallamaram	881.90	792.10	1075.54	3036.90	686.75	629.82
		7.43	*6.68*	*9.06*	*25.59*	*5.79*	*5.31*
7	Silarpatti	1279.37	1017.53	775.96	2332.75	875.62	1226.39
		9.71	*7.72*	*5.89*	*17.71*	*6.65*	*9.31*

(Contd.)

(Contd. Table 6.2)

SlNo	Village	Fertiliser	Pesticide/ Insecticide	Herbicides	Electricity	Machinery	Total
1	Pulkattai	644.40	141.02	94.24	645.71	838.73	7270.29
		8.86	*1.94*	*1.30*	*8.88*	*11.54*	*100*
2	Vannivelampatti	553.01	370.53	116.58	1413.32	1693.64	11227.08
		4.93	*3.30*	*1.04*	*12.59*	*15.09*	*100*
3	Kunnathur	786.88	391.37	194.25	1218.14	1870.73	11359.85
		6.93	3.45	1.71	10.72	16.47	100
4	Velampur	701.79	741.56	215.79	1289.10	2793.66	14295.37
		4.91	*5.19*	*1.51*	*9.02*	*19.54*	*100*
5	Genjampatti	854.60	665.79	161.84	760.09	2460.75	14308.31
		5.97	*4.65*	*1.13*	*5.31*	*17.20*	*100*
6	Nallamaram	1033.75	688.48	156.66	867.15	2016.71	11865.75
		8.71	*5.80*	*1.32*	*7.31*	*17.00*	*100*
7	Silarpatti	1063.20	873.55	145.00	1879.08	1704.83	13173.29
		8.07	*6.63*	*1.10*	*14.26*	*12.94*	*100*

(Contd.)

(Contd. Table 6.2)

SlNo	Village	Human Energy	Animal Energy	Seed Input	FYM	Wood Instrument	Diesel / Petrol
8	Shandaiyur	671.57	841.45	615.00	3377.75	400.64	705.25
		5.90	*7.40*	*5.41*	*29.70*	*3.52*	*6.20*
9	Melapatti	528.90	669.40	651.09	3799.28	348.96	893.54
		5.22	*6.61*	*6.43*	*37.53*	*3.45*	*8.83*
10	Pappaiyapuram	871.35	714.57	735.61	3328.80	696.89	583.88
		8.18	*6.70*	*6.90*	*31.23*	*6.54*	*5.48*
11	Kuvalapuram	902.85	822.24	1440.10	2772.10	744.56	894.59
		6.80	*6.19*	*10.84*	*20.88*	*5.61*	*6.74*
12	Chittulotti	800.92	547.39	1367.16	3053.56	539.97	404.43
		7.37	*5.04*	*12.59*	*28.11*	*4.97*	*3.72*
	Average	744.66	782.20	1154.99	3132.06	575.93	676.71
		6.53	*6.86*	*10.13*	*27.47*	*5.05*	*5.94*

(Contd.)

(Contd. Table 6.2)

SlNo	*Village*	*Fertiliser*	*Pesticide/ Insecticide*	*Herbicides*	*Electricity*	*Machinery*	*Total*
8	Shandaiyur	705.03	896.67	211.81	704.67	2243.58	11373.43
		6.20	*7.88*	*1.86*	*6.20*	*19.73*	*100*
9	Melapatti	751.91	433.79	63.85	445.41	1538.52	10124.64
		7.43	*4.28*	*0.63*	*4.40*	*15.20*	*100*
10	Pappaiyapuram	748.89	713.38	116.97	995.02	1152.82	10658.16
		7.03	*6.69*	*1.10*	*9.34*	*10.82*	*100*
11	Kuvalapuram	669.31	735.44	233.42	1586.17	2478.35	13279.14
		5.04	*5.54*	*1.76*	*11.94*	*18.66*	*100*
12	Chittulotti	685.68	686.00	210.15	599.37	1966.63	10861.22
		6.31	*6.32*	*1.93*	*5.52*	*18.11*	*100*
	Average	752.60	579.15	156.34	994.06	1852.88	11401.57
		6.60	*5.08*	*1.37*	*8.72*	*16.25*	*100*

Table 6.3

Source-wise Energy Consumption (kWh/acre) by Selected Irrigated Crops

Crops	*Human Energy*	*Animal Energy*	*Seed Input*	*FYM*	*Wood Instrument*	*Diesel / Petrol*
Bullock Operated Farm						
Paddy	1836.36	3343.15	207.12	12884.58	1555.95	603.42
	5.91	*10.76*	*0.67*	*41.48*	*5.01*	*1.94*
Sugarcane	1295.74	1274.10	24640.69	10675.00	589.79	2396.70
	2.52	*2.48*	*47.99*	*20.79*	*1.15*	*4.67*
Cotton	2454.62	2720.25	102.29	12422.50	1550.17	330.06
	9.03	*10.01*	*0.38*	*45.72*	*5.70*	*1.21*
Groundnut	1029.07	1549.10	1090.03	12033.27	1307.42	143.88
	5.36	*8.07*	*5.68*	*62.66*	*6.81*	*0.75*
Sorghum	1130.00	1676.18	60.35	11506.25	773.94	122.88
	5.91	*8.76*	*0.32*	*60.16*	*4.05*	*0.64*
Bajra	1208.40	1636.80	23.74	11025.00	593.04	2.78
	6.83	*9.25*	*0.13*	*62.32*	*3.35*	*0.02*
Pulses	1062.66	1129.94	113.90	11137.05	659.50	0.00
	6.79	*7.22*	*0.73*	*71.13*	*4.21*	*0.00*

(Contd.)

(Contd. Table 6.3)

Crops	*Fertiliser*	*Pesticide / Insecticide*	*Herbicides*	*Electricity*	*Machinery*	*Total*
Bullock Operated Farm						
Paddy	4957.23	462.92	501.70	3774.18	933.38	31059.98
	15.96	*1.49*	*1.62*	*12.15*	*3.01*	*100*
Sugarcane	7116.14	68.39	326.25	1506.83	1460.30	51349.92
	13.86	*0.13*	*0.64*	*2.93*	*2.84*	*100*
Cotton	5137.00	1020.52	72.14	865.80	497.48	27172.82
	18.90	*3.76*	*0.27*	*3.19*	*1.83*	*100*
Groundnut	970.25	368.23	24.17	457.73	230.82	19203.95
	5.05	*1.92*	*0.13*	*2.38*	*1.20*	*100*
Sorghum	2293.64	318.30	6.04	978.03	261.54	19127.14
	11.99	*1.66*	*0.03*	*5.11*	*1.37*	*100*
Bajra	1470.22	1047.84	6.04	560.55	117.15	17691.56
	8.31	*5.92*	*0.03*	*3.17*	*0.66*	*100*
Pulses	56.10	1002.60	0.00	433.65	61.36	15656.77
	0.36	*6.40*	*0.00*	*2.77*	*0.39*	*100*

(Contd.)

(Contd. Table 6.3)

Crops	*Human Energy*	*Animal Energy*	*Seed Input*	*FYM*	*Wood Instrument*	*Diesel / Petrol*
Tractor Operated Farm						
Paddy	1530.01	66.68	207.20	12896.55	0.00	1661.63
	5.47	*0.24*	*0.74*	*46.09*	*0.00*	*5.94*
Sugarcane	1175.97	0.00	26265.76	10675.00	0.00	2745.77
	2.26	*0.00*	*50.47*	*20.51*	*0.00*	*5.28*
Cotton	2360.74	0.00	96.73	12422.50	0.00	5764.04
	7.14	*0.00*	*0.29*	*37.58*	*0.00*	*17.44*
Groundnut	909.66	0.00	1090.03	12163.83	0.00	1293.82
	4.99	*0.00*	*5.98*	*66.72*	*0.00*	*7.10*
Sorghum	971.09	0.00	60.35	11506.25	0.00	1442.80
	5.19	*0.00*	*0.32*	*61.51*	*0.00*	*7.71*
Bajra	1146.12	0.00	23.74	11025.00	0.00	1091.22
	6.65	*0.00*	*0.14*	*63.97*	*0.00*	*6.33*
Pulses	878.40	0.00	113.91	11137.05	0.00	856.29
	5.73	*0.00*	*0.74*	*72.68*	*0.00*	*5.59*

(Contd.)

(Contd. Table 6.3)

Crops	*Fertiliser*	*Pesticide / Insecticide*	*Herbicides*	*Electricity*	*Machinery*	*Total*
Tractor Operated Farm						
Paddy	5037.59	462.92	627.50	3800.79	1688.11	27978.98
	18.00	*1.65*	*2.24*	*13.58*	*6.03*	*100*
Sugarcane	7121.39	68.39	314.17	1506.83	2167.99	52041.25
	13.68	*0.13*	*0.60*	*2.90*	*4.17*	*100*
Cotton	5149.42	1070.52	94.51	878.30	5219.20	33055.96
	15.58	*3.24*	*0.29*	*2.66*	*15.79*	*100*
Groundnut	991.08	368.23	24.17	458.22	931.50	18230.55
	5.44	*2.02*	*0.13*	*2.51*	*5.11*	*100*
Sorghum	2293.64	380.88	6.04	979.11	1066.59	18706.75
	12.26	*2.04*	*0.03*	*5.23*	*5.70*	*100*
Bajra	1478.41	1057.15	6.04	560.55	847.24	17235.49
	8.58	*6.13*	*0.04*	*3.25*	*4.92*	*100*
Pulses	56.10	1002.60	0.00	537.68	742.23	15324.26
	0.37	*6.54*	*0.00*	*3.51*	*4.84*	*100*

Table 6.4
Source-wise Energy Consumption (kWh/acre) by Selected Unirrigated Crops

Crops	*Human Energy*	*Animal Energy*	*Seed Input*	*FYM*	*Wood Instrument*	*Diesel / Petrol*
Bullock Operated Farm						
Cotton	1727.95	1512.74	114.06	9402.20	1255.25	132.65
	10.85	*9.50*	*0.72*	*59.06*	*7.88*	*0.83*
Groundnut	844.18	2066.96	1280.61	11060.35	1761.45	91.85
	4.71	*11.54*	*7.15*	*61.75*	*9.83*	*0.51*
Sorghum	943.34	1720.50	93.08	10456.25	686.14	144.74
	6.14	*11.20*	*0.61*	*68.05*	*4.47*	*0.94*
Bajra	919.82	1929.56	32.64	10587.50	770.66	4.18
	5.30	*11.11*	*0.19*	*60.99*	*4.44*	*0.02*
Pulses	962.63	1173.25	141.53	10141.67	657.79	0.00
	6.80	*8.29*	*1.00*	*71.67*	*4.65*	*0.00*

(Contd.)

(Contd. Table 6.4)

Crops	*Fertiliser*	*Pesticide / Insecticide*	*Herbicides*	*Electricity*	*Machinery*	*Total*
Bullock Operated Farm						
Cotton	1178.20	441.88	24.17	0.00	132.00	15921.08
	7.40	*2.78*	*0.15*	*0.00*	*0.83*	*100*
Groundnut	432.47	231.46	24.17	0.00	118.64	17912.12
	2.41	*1.29*	*0.13*	*0.00*	*0.66*	*100*
Sorghum	670.33	378.75	120.83	0.00	152.63	15366.58
	4.36	*2.46*	*0.79*	*0.00*	*0.99*	*100*
Bajra	1247.98	1855.54	6.04	0.00	6.19	17360.11
	7.19	*10.69*	*0.03*	*0.00*	*0.04*	*100*
Pulses	29.30	1044.49	0.00	0.00	0.00	14150.65
	0.21	*7.38*	*0.00*	*0.00*	*0.00*	*100*

(Contd.)

(Contd. Table 6.4)

Crops	*Fertiliser*	*Pesticide / Insecticide*	*Herbicides*	*Electricity*	*Machinery*	*Total*
Tractor Operated Farm						
Cotton	1178.19	441.88	24.17	0.00	2108.24	16313.01
	7.22	*2.71*	*0.15*	*0.00*	*12.92*	*100*
Groundnut	428.72	231.47	24.17	0.00	726.04	15609.41
	2.75	*1.48*	*0.15*	*0.00*	*4.65*	*100*
Sorghum	717.89	378.75	120.83	0.00	849.75	14856.76
	4.83	*2.55*	*0.81*	*0.00*	*5.72*	*100*
Bajra	1247.98	1855.54	6.04	0.00	765.08	16566.52
	7.53	*11.20*	*0.04*	*0.00*	*4.62*	*100*
Pulses	29.30	1044.49	0.00	0.00	533.47	13620.82
	0.22	*7.67*	*0.00*	*0.00*	*3.92*	*100*

(Contd.)

Table 6.5

REGRESSION CO-EFFICIENTS OF ENERGY CONSUMPTION MODELS (Crop-wise) IN IRRIGATED FARM

Variable	*Paddy*	*Sugarcane*	*Cotton*	*Groundnut*	*Sorghum*	*Bajra*	*Pulses*
Area	4181.38[1]	87336.94[1]	4062.73[1]	1911.65[1]	10003.1[1]	4993.69[1]	1796.50[1]
Animate Energy	0.10048[2]	-7.47312[1]	0.01282	0.01799	-0.833[1]	-0.29799[2]	0.02579
Bio-Energy	--	-0.18837	0.095[1]	--	0.1895[2]	0.40885[1]	--
Mech. Energy	0.1912[1]	0.14072	0.0303[2]	0.00203	--	-0.08543	--
Chem. Energy	--	--	--	0.38224[1]	--	--	0.26338
Intercept	-316.58[1]	6069.01[2]	-739.37[1]	-96.34[2]	-747.60[2]	-0.499.3[2]	-31.66
F Value	9926.98[1]	1083.48[1]	1909.87[1]	13886.2[1]	6499.69[1]	1423.3[2]	852.98[1]
R^2 Value	0.99863	0.99335	0.99583	0.99975	0.99928	0.99825	0.99688
No.of Cases (60)	9	5	12	6	11	10	7

Y = Yield in kg. [1] = Significant at p = .001
[2] = Significant at p = .01

Note:
a) Animate energy includes human and bullock labour involved in agricultural activities.
b) Bio-energy includes seed input, FYM, green manure and wood instruments.
c) Mechanical energy includes tractor, power tiller, thresher, sprayer, motor, diesel, petrol, oil, and electricity
d) Chemical energy includes fertilisers, pesticieds, insecticides, herbicides and fungicides used.

Table 6.6
REGRESSION CO-EFFICIENTS OF ENERGY CONSUMPTION MODELS (Crop-wise) IN UNIRRIGATED FARM

Variable	*Cotton*	*Groundnut*	*Sorghum*	*Bajra*	*Pulses*
Area	5708.71[1]	2500.51[1]	8179.77[1]	2110.56[1]	1618.21[1]
Animate Energy	-34503[1]	0.03869	0.000244	-2347	0.05648[1]
Bio-Energy	0.4006[2]	-13750	--	--	-0.00067
Mech. Energy	--	--	0.04377	--	--
Chem. Energy	-0.22601	--	1.591[1]	0.99541[2]	--
Intercept	-549.59[2]	-611.66[1]	-639.77[1]	-1242.98[2]	-9.29
R^2 Value	0.99894	0.99790	0.99960	0.98936	0.99971
No.of Cases (180)	46	21	59	22	32

Y = Yield in kg. [1] = Significant at $p = .001$
[2] = Significant at $p = .01$

Table 6.7
REGRESSION CO-EFFICIENTS OF ENERGY CONSUMPTION MODELS (Operation-wise) IN IRRIGATED FARM

Variable	*Bullock Operated Farm*	*Tractor Operated Farm*
Area	-23479.33[1]	-35215.57[1]
Animate Energy	-0.997[2]	--
Bio-Energy	2.972[1]	5.031[1]
Mech. Energy	0.24002	0.043
Chem. Energy	2.044[2]	- 1.816[2]
Intercept	-6694.3[2]	-7241.19
F Value	66.575[1]	210.047[1]
R^2 Value	0.78163	0.96552
No.of Cases (60)	37	23

Y = Yield in kg. [1] = Significant at p = .001
[2] = Significant at p = .01

Table 6.8
REGRESSION CO-EFFICIENTS OF ENERGY CONSUMPTION MODELS (Operation-wise) IN UNIRRIGATED FARM:

Variable	*Bullock Operated Farm*	*Tractor Operated Farm*
Area	532.989	1323.564
Animate Energy	0.2809[2]	-3.506
Bio-Energy	-0.1878	--
Mech. Energy	17.937[1]	2.731[1]
Chem. Energy	1.168[2]	0.4323
Intercept	2964.83[2]	-1105.67
F Value	16.42[1]	23.48[1]
R^2 Value	0.61682	0.78288
No.of Cases (180)	74	106

Y = Yield in kg. [1] = Significant at p = .001
[2] = Significant at p = .01

Table 6.9
REGRESSION CO-EFFICIENTS OF ENERGY CONSUMPTION MODELS (Category-wise) IN IRRIGATED FARM:

Variable	*Marginal Farmer*	*Small Farmer*	*Medium Farmer*	*Large Farmer*
Area	8368.93	-11778.57	-27864.7[1]	-35020.52[1]
Animate Energy	-10.7899[1]	-8.6703[1]	-0.1054	-5.077[1]
Bio-Energy	1.1426[1]	2.7637[1]	3.4306[1]	5.3606[1]
Mech. Energy	-2.3193[2]	-0.3598	0.0401	0.434
Chem. Energy	8.6472[1]	5.2918[1]	-0.0749	-2.0889[2]
Intercept	-3821.68	-6173.11	1635.016	-3271.46
F Value	71.968[1]	68.727[1]	59.539[1]	304.251[1]
R^2 Value	0.85505	0.87520	0.92253	0.98638
No.of Cases (60)	4	9	42	5

Y = Yield in kg. [1] = Significant at p = .001
[2] = Significant at p = .01

Table 6.10
REGRESSION CO-EFFICIENTS OF ENERGY CONSUMPTION MODELS (Category-wise) IN UNIRRIGATED FARM

Variable	*Marginal Farmer*	*Small Farmer*	*Medium Farmer*	*Large Farmer*
Area	19183.939[1]	7559.451[1]	10223.809[1]	-627.618
Animate Energy	2.8165[2]	0.43798[2]	1.5618	0.7395
Bio-Energy	-1.9024[1]	-0.31060	-1.2111	--
Mech. Energy	0.66512	-1.2636	1.4356	2023[2]
Chem. Energy	-3.5225[1]	-0.13169	0.63915	0.4525
Intercept	-851.9156	-5919.75	2542.36	2621.956
F Value	8.1203[1]	8.2089[1]	1.869	9.684[1]
R^2 Value	0.55164	0.67237	0.27216	0.67091
No.of Cases (180)	65	60	37	18

Y = Yield in kg. [1] = Significant at p = .001
[2] = Significant at p = .01

Index

A

Adi Dravidira, 96
Administration, 95
Agricultural
  Activities, 76
  Based Countries, 57
  Development, 26, 123
  Dryland, 24
  Ecological, 82
  Economy, 149
  Energy (ies), 48, 70
  Energy Consumption, 142, 144
  Energy Intensive, 65
  Engineering, 165
  Farms, 26
  Implements and Fertilizers, 23
  Machinery and Implements, 104
  Machinery, 54
  Modern, 23, 82
  Operations, 144
  Practices, 63, 69
  Products, 23
  Production, 24
  Research Institute Madurai, 75
  System,
  Modern, 60
  Traditional, 60
  Technological Development—, 26
  Technology, 27, 55
  Traditional, 51, 82, 114
Agriculture Energy
  Consumption, 158, 159
Agro-
  Based Industries, 34
  Chemicals, 88, 122, 164
  Economic System, 42

Ecosystem, 31, 34, 36, 63, 114, 168
Forestry Program, 165
Alleviation of Poverty, 27
Alternative Policies, 166
Animal
Energy, 57, 133, 142, 143, 146, 147, 150, 151, 152, 160
Machine System, 169
Population, 143
Power, 170
Artificial Crop Drying, 53

B

Bio-
Energy, 127, 128, 148, 150, 151, 153, 155, 162, 168
Fertilizers, 75, 148
Biological
Control of Pests, 161
Pest Control Methods, 118, 119
Black Cotton, 122
Block,
Kalligudi, 91
Peraiyur, 97
Sivakasi and Watrap, 91
T. Kalluppatti, 92, 93, 95, 102, 141, 142, 143, 144, 158
Tirumangalam, 91
Bullock Operated Farms (BOF), 113, 115, 116, 118, 122, 124, 125, 129, 132, 146-148, 154, 160

C

Canadian Agriculture,
Productive Measures in-, 67
Canals and Dams, 55
Cash Crops, 23
Sugarcane—, 24
Chemical
And Mechanical Energy, 145
Energy, 51, 61
Fertilizers, 52, 59, 119, 161, 162
China, 112
China and India, 63
Chinese Wet Rice Cultivation, 44
Commercial
Crops, 24
Energy, 34, 36, 49, 50, 81, 145, 148
Compost, 83
Cotton
Farmers, 52
In Gujarat, 52
Crop
Production, 150
Production System, 28
Rotation, 53
Cropping
Pattern, 24, 25, 84, 151, 168
System, 165
Crop-Wise Energy
Consumption, 112

D

District
- Bareilly (UP), 48
- Coimbatore (TN), 58
- Hissar (Haryana), 58
- Madurai (Tamil Nadu), 24, 33, 55, 70-72, 93, 94, 143
- Rural Development Agency, (DRDA), 95
- Virudhnagar, 91, 94

Draught
- Animals, 24, 83, 142, 143
- Animals in India, 56, 57
- Prone Areas, 167

Dry
- Crops, 160
- Farming, 124, 129
- Farming Cotton, 125

E

Eco-Friendly Energy, 63
Ecology and Agro-Environmental, 152
Economic
- Crops 82
- Friendly Energy Resources, 36
- System, 28

Economy, 157
- Agricultural, 149

Ecosystem, 85
Educational Institutions, 94
Egypt, 112
Electrical Energy Consumption, 147
Electromagnetic Radiation, 61
Energy, 26
- And Agriculture, 39, 142
- Bio-mass, 26
- Chemical, 51, 61, 152, 154, 155
- Fertilizers, 43
- Classification, 71-90
- Commercial, 24, 25, 34, 36, 79, 114, 115, 117, 119, 124, 133, 159, 162
- Commercial/Non-Commercial, 111
- Conservation, 67
- Conservation Measures, 156
- Consumption, 25, 76, 78, 112, 125, 126, 145, 146, 148, 149, 153, 159, 164
- For HYV Paddy, 54
- In Dry Farms, 147, 148
- Model, 141
- Pattern, 46, 163, 167
- Pattern in India, 49
- Defined, 78

Efficient Technologies, 27
Embodied, 80
- Fossil, 26
- Gross, 80
- Intensive Agriculture, 65
- Management, 165
- Management in Agriculture, 164
- Mechanical, 151, 153, 155, 159, 160, 163

Non-Commercial, 114, 119, 124, 160, 168

Energy, 79

Nuclear, 40, 41

Planning, 33

Primary, 79

Productivity, 44

Renewable Sources of-, 165

Requirement of Agriculture in-, (ERA), 46

China, 46

India, 46

Japan, 46

Taiwan, 46

USA, 46

Secondary, 79

Solar, 23, 26, 60-63, 146

Subsidies, 169

System, 157

Technology, 169

Use Pattern, 111

In Agriculture, 141

Units, 76

Useful Energy, 79

Wind, 26

Environment/Assessment Impact, (EIA), 166

Environmental Degradation, 26

F

FAO Statistics, 46

Farm

Activities, 71

Energy, 43

Forestry, 165

Labourers, 27

Machinery Management, 165

Machinery, 50

Mechanization, 25

Operations, 145, 149

Size, 48

Yard Manure, (FYM), 28, 39, 52, 58, 59, 76, 83, 88, 112, 115, 119, 129, 133, 142, 144, 145-147, 159, 161

Energy, 144, 148, 160

Farmers Classification, 24

Large, 24, 70, 75, 129, 134, 135

Marginal, 24, 70, 75, 129, 133, 161

Medium, 24, 70, 75, 139

Small, 24, 70, 75

Farmers,

Small and Marginal, 165

Small, Medium, Large, Marginal, 106, 113

Farming, 53

Categories, 69

In Nilgiri, 53

Operations, 124, 129

Practices, 29, 152, 162

System, 35, 53

Techniques, 36

Farms, Small, 121, 122

Fertilizers

And Pesticides, 24, 31, 34, 43, 49, 50, 59, 64, 67, 79, 81, 115, 117

Energy, 146, 148

Food
Production System, 45
System, 45, 66
Fossil
Energy, 26, 31
Fuels, 23, 40, 60, 65
Fuel Based Fertilizers, 23
Hydrocarbon, 41
Fuels Classification, 80

G

Genjamatti Village Panchayat, 142, 143, 159
Global Radiation, 61
Gram Panchayats, 59
Green
Manuring, 59, 83
Revolution, 55, 129
Revolution in India, 25
Groundnut Production, 129
Groundwater Sources, 56

H

Haryana, 25
Herbicides, 142
High Technology Agricultural Technologies, 31
High Yielding Varieties (HYV), 25, 47, 57
Of Seeds, 55

I

India's
Energy Economy, 26
Farming, 57
Indian
Agriculture, 25, 27, 28, 49
Farms, 67
Research Institute, Kerala, 47
System, 47, 50
Industrialized Agriculture, 54
Production System, 54
Insecticides, 143
Integrated Energy Management, (IEM), 168
Intensive Agricultural Production, 43
Irrigated Farming, 56
Irrigated/Rain-fed Farms, 149
Irrigation, 55, 56, 64, 67, 87, 112, 113, 118, 122, 123, 125
And Threshing, 48
Energy, 53
Farm, 145
Farming System, 55
Methods, 95, 103
Minor, 95
Surface, 95
Pumping in India, 56
Sources, 102
Italy, 112

J

Japan, 112

K

Kunnathur Village Panchayat, 142, 158

L

Land
- Classification, 92
- Holdings, 106, 141
- Large Farmers, (LF), 113, 116
- Lift Irrigation, 56
- Live Stock Rearing, 71

M

Madurai City, 91
Manure Energy, 147
Manuring and Irrigation, 84
Marginal Farmers (MF), 113, 125, 127, 128 see under Farmers also
Mechanical (ized)
- Energy, 151, 153, 155, 159, 160, 163
- Power, 43, 150

Medium Farmers (MeF), 113, 117
Melappati and Silarpatti Panchayats, 143
Modern Technologies of Farming, 31, 32
Mono Crop to Multi Crop, 34

N

National Commission of Agriculture, 28, 134
Natural Gas Prices, 65
New Irrigation Sources, 55
Non-Commercial
- Energy (NCE), 28, 41, 57, 81, 168
- Fuels, 34

Non-
- Energy Technologies, 32
- Renewable Energy (NRE), 24, 27, 67
- Renewable Oil Stocks, 32
- Renewable Sources, 23

Nuclear Energy, 40, 41

O

Oil
- Based/Electricity Based Machines, 43
- Crops and Pulses, 50

Operational Skills and Techniques, 53
Organic Manure, 63, 122

P

Panchayat Union, 875
Participatory Rural Approach, (PRA), 75
Pesticides, 24
Pesticides Application, 143, 145
Photosynthesis Process, 61
Photo-Synthetically Active
- Radiation, (PAR), 61, 62
- Techniques, 75

Planned Agricultural Development, 118
Plant Protection Chemicals, 89

Policy Issues, 65
Population Growth, 25, 28
Potato Production, 54
Processing Industry, 44
Production Technology, 67
Pulkattai Village Panchayat, 142, 158
Punjab, 25
    Agricultural University, 43
    Area, 53

R

Regression Model, 162
Renewable
    Sources of Energy, 165
Renewable/Non-renewable
    Resources, 40
    Technologies, 31
Rice-Wheat Cropping System, 59
Rotary Applications, 170
Rotation Crops, 84
Rural
    Communication of India, 54
    Development Program, 95, 166
    Electrification, 55
    Energy Program, 166
    Energy System, 166
    Sectors, 25

S

Sadhugiri Hills, 91
Sandhaiyur Panchayat, 143, 144
Sector,
    Agriculture, 25, 26, 27, 54, 59, 129
    Industrial, 26, 95
Seed Energy, 145
Short Duration Crops, 54
Small
    And Marginal Farmers, 165 see under Farmers also
    Farmers, 127, 128
Social Forestry, 165
Socio-Economic
    Structure, 31
    System, 31
Solar
    Dry Farming, 26
    Energy, 23, 60-63, 146
    Radiation, 33, 40, 60, 62
Subsistence Farming System, 31
Sugarcane, 118
    Cultivation, 146, 147
    Production, 119, 120, 122
Sustainable Development, 85

T

T Kallappatti Panchayat Union, 95
Tamil Nadu, 25
Technologies of Production, 31
Technology, 157
    Of Energy Conversion, 41
TN Agricultural University, 75
Toxic Agriculture and
    Chemicals, 51

Tractor Operated Farms, (TOF), 54, 113-116, 119, 126, 129, 133-135, 139, 146-148, 154, 160
Traditional Agricultural in Nicaragua, 65
Transfer of Technologies to Farmers, 165

U

United Kingdom, 44
Urbanization, 26
US, 112
  Agriculture, 31
  And UK, 53
  Food System, Energy Used in-, 44
  Maiz Products, 24
  Western States, 56

V

Velampur Panchayat, 142-144
Village
  Administrative Officer (VAO), 97
  Panchayats Classification, 93
  Panchayats for Study, (VPS), 73
  Chitulotti, 73
  Ganjampatti, 73
  Koovalapuram, 73
  Kunnathur, 73
  Melapatti, 73
  Nallamaru, 73
  Pappaiyapundu, 73
  Pulkattai, 73
  Samdhaiyev, 73
  Silarpatti, 73
Vannivelampatti, 73
Velampur, 73
  Panchayats, 141, 158

W

Water
  Lifting, 170
  Resources, 167
West Uttar Pradesh, 25
Western Ghats, 91